D1548398

First published as *How to Build and Repair Fences and Gates* in 2007 by Voyageur Press, an imprint of MBI Publishing Company, 400 First Avenue North, Suite 400, Minneapolis, MN 55401 USA.
This edition published 2014.

Library of Congress Cataloging-in-Publication Data

Kubik, Rick, 1953-
[How to build & repair fences & gates]
Farm fences and gates : build and repair fences to keep livestock in and pests out / by Rick Kubik.
pages cm
Revised editon of: How to build & repair fences & gates.
Includes index.
ISBN 978-0-7603-4569-6 (softcover)
1. Fences--Design and construction. 2. Fences--Maintenance and repair. 3. Farm equipment. I. Title.
TH4965.K824 2014
631.2'7--dc23

2013033563

Editors: Amy Glaser, Elizabeth Noll, and Madeleine Vasaly
Design Manager: Cindy Samargia Laun
Designer: Kim Winscher

Front cover: Chris Lofty/Dreamstime.com
Interior photographs are from the author, except as noted: frontis: David Handley/Getty Images; TOC: Gay Bumgarner/Alamy; Page 7: Lightpoet/Dreamstime.com; Page 8: Marussia/Shutterstock.com; Page 10: Spirit of America/Shutterstock.com; Page 14: Carlos Caetano/Shutterstock.com; Page 17: Dariusz Gora/Shutterstock.com; Page 54: Sanddebeautheil/Shutterstock.com; Page 67: David Handley/Getty Images; Page 68: Regis Cordeiro/Shutterstock.com; Page 128: Tsekhmister/Shutterstock.com; Page 131: TOMO/Shutterstock.com; Page 133: Reinout Van Wagtendonk/Dreamstime.com; Page 139: iStockphoto.com/Robas; Page 144: BGSmith/Shutterstock.com; Page 145: Wild Art/Shutterstock.com; Page 147: Debbie Steinhausser/Shutterstock.com; Page 149: Schmaelterphoto/Shutterstock.com; Page 162: F1online digitale Bildagentur GmbH/Alamy; Page 163: Judy Kennamer/Shutterstock.com; Page 164: J. Marijs/Shutterstock.com

Printed in China
10 9 8 7 6 5 4 3 2 1

FARM FENCES & GATES

Build and Repair Fences to Keep Livestock In and Pests Out

Rick Kubik

CONTENTS

INTRODUCTION

Today's fences are now key elements in protecting modern smallholding livestock such as chickens, goats, and many others, along with protecting smallholding gardens. In addition, well-designed fences protect smallholdings against wild animals such as deer, bear, and raccoons, and many others. What's more, fences help protect your smallholding against inadvertent human traffic such as passersby and their car and truck traffic.

Keeping livestock on their proper pastures and away from other crops remains a primary function of fences, but a critical secondary function is keeping predators and pest animals away from livestock and crops. When you're planning a fence that won't fall down in the face of marauding raccoons or deer, it's useful to remember the principles of building a fence that won't fall down in the face of livestock such as cows or hens. When you need to keep something out, use the same principles that you use to keep things in.

An equally important function in these times of increased traffic speed and density is to keep livestock and wild animals off roads where they could get involved in traffic accidents that can be fatal to increasing numbers of both animals and motorists. Many motorists or motorcyclists grump about fences that keep them out of farm lands, but from the farmers' point of view, it's more a case of keeping heavy, awkward livestock off the roads where fast-moving vehicles cause all sorts of carnage to humans, vehicles and animals. To support this point, check in your area for stories of what happens when a car hits a deer, or a pig rolls under vehicles, or vehicles skid on road kill. The stories will not be happy, but they will illustrate the need to fence vehicles away from slow-moving smallholder livestock. There are currently more fencing systems than ever before to choose from when confining or excluding various kinds and sizes of livestock. Until recently, barbed wire was the first and often only choice of agricultural fence for anything but short runs. Now the maturation of high-tensile wire and electrified fence technology has added new choices while reducing costs and increasing effectiveness. In certain cases, the proliferation of welded steel panel, plastic, and premade wooden rail fences has added new farm and yard fencing options at reasonable cost.

Whether you're new to building farm fences or are more experienced and looking for better ways to build, this book provides practical information on why specific fences may fit your needs. Although most types of fencing have many applications on the farm, this book provides as many specifics as possible to help you make the decision on the one type that is best suited to the specific function you have in mind. Once the kind of fence is chosen, plenty of how-to tips help you get it built in a way that makes the most efficient use of your money, time, and labor, while also keeping you and your fencing crew safe from injury.

The specific functions considered cover the most common types of livestock (horses, cattle, sheep, goats, pigs, and poultry) and predators or pests (coyotes, foxes, and deer). But the information also takes into account some of the more exotic problems rural dwellers are trying to solve, such as fencing in bison or keeping out bears, wild geese, and roving packs of domestic dogs.

Information concentrates on the most widely used types of fencing (high-tensile wire, electrified, and barbed-wire) and explains why their advantages in cost and effectiveness have led them to become top choices. It also provides a review of some of the older methods, such as split-rail, that are still quite effective and may offer important advantages where cash is in short supply but wood and labor are not. The book also looks at options for wooden fence posts that comply with organic production standards.

Fences can do more than just manage the movement of livestock, which is why this book also takes a look at what fences can do for managing environmental factors, such as wind, water, drifting snow, wildlife habitat, and all-around farm beautification. The right type of fence for environmental management can go a long way toward making your farm a more pleasant and sustainable place to live and work.

Any type of fence requires a considerable investment in planning, materials, time, and labor to build. Since the costs skyrocket if you have to take down and redo an incorrectly located fence, the book starts off with a practical discussion of how to make sure the fence is in the right place. Property law and fence law can be a minefield for the unwary, so it's wise to get a good feel for the issues before you ever dig a hole or drive a staple.

A huge thank you is due to the people who allowed me onto their farms to take

A good fence keeps the chickens in and the predators out.

Animals are always testing the fences in their way. It's an ongoing challenge to find less-costly, more-effective forms of fencing to keep livestock safe from harm.

many of the photos that illustrate this book. Special thanks to those who allowed me to take the photo sequences of work as it progressed: Bill and Vera Mokoski, retired operators of the Treco Ranch, and their neighbors Claude and Debbie Delisle, and children Hannah, Naomi, and Jaclyn, who decided to take active steps toward living on a new dream small farm. Thanks also to the many agricultural extension agents across this continent who work hard at freely providing agricultural information, including fence building. Their work has created a priceless wealth of hard-nosed, unbiased knowledge for those who chose to live and work on the land, including the growing ranks of new small farmers. I would strongly encourage you to explore the wide range of agricultural information available from your state or provincial agricultural extension department. Thanks in particular to Lance Brown, recently retired from the British Columbia Ministry of Agriculture, Food, and Fisheries, who provided some dramatic photos of wooden rail fencing in this book, and who during his career prepared several papers referenced in this book on various aspects of fencing.

My own experience with building and repairing fences came about when growing up on a farm. When first faced with sitting down to write this book, I soon became concerned about that experience being enough to make a whole book. But as I reviewed the things I had taken for granted that people would know, and looked to the work of others mentioned above for a broader overview of the topic, the material began to come together. The process has provided me with a much better understanding of why things were done the way they were and why things have changed to make fencing more effective and efficient. I trust you'll find that kind of in-depth information communicated effectively enough that you can put it to use improving your own knowledge of fences and how to build them.

Rick Kubik
former certified crop adviser
American Society of Agronomy

CHAPTER

1

START IN THE RIGHT PLACE

Fences built in the right place mean less time and money spent rebuilding or removing them and fewer legal and environmental problems.

Before you start gathering the materials and information you'll need to build any kind of fence, be sure you've got reliable information on where to build it.

That may seem pointless and needlessly expensive when low costs, getting off the grid, back to the land, and living on a shoestring budget are top of mind for you. But getting the fence done once and in the right place is a lot less costly and less bothersome than having to argue with neighbors about where the fence is, and then either removing and redoing the fence or having an angry, long-running feud over fence placement. Even the simplest fence involves as much or more time and work to take down and move than it does to put up in the first place. Making sure the fence is in the right place to begin with can save you from having to tear it down and do it all over again, along with the frustration, cost of materials, time, labor, bad feelings with neighbors, and potential for litigation that involves. Whatever else a fence needs to keep in or out, for your own sake it also needs to keep lawyers out!

For fences known to be completely within the boundaries of your property, determining the right location is relatively simple. It mostly involves assessing land features, such as hills and watercourses, whether there is a suitable path to bring in a tractor and post pounder, or whether the soil is suitable for securely holding posts.

It's when you are at or near the property boundaries you share with neighbors, highways, or other public areas like forests and lakes that the issue of correct location becomes extremely important and potentially vexing. Every state and province has quite an extensive body of fence law developed in recognition of all the disputes that have arisen over the years.

Any boundary fence (also known as line fence or partition fence) may involve dealing with issues other than placing the fence on a straight line between property corners. Along with the correct alignment, there is the issue of whether a neighbor should pay part of the cost of a fence you put up between your properties. That's why in these cases, it's highly recommended to talk with neighboring landowners and get a written agreement and/or consult a land surveyor professionally qualified in your jurisdiction.

It's commonly assumed that locating your property boundaries is simply a matter of stringing a straight line between the corner posts or setting off from a neighbor's existing fence corners. If the previous layout of your property or adjoining land parcels were done recently and correctly, that will most likely be true. However, many readers of this book are taking on small farms in more remote rural areas where surveys were done long ago, if at all, and aren't necessarily correct.

> **. . . experience as a surveyor to a land titles office has shown me that whenever a subsequent and carefully made survey of land abutting on a railway right of way has been made, and a plan of same presented for registration, fully half of the land ties shown on the railway right of way plan have been found to be incorrect—often grossly so.**
>
> ***Descriptions of Land: A Text-Book For Survey Students*** **by R.W. Cautley, D.L.S., B.C.L.S, A.L.S.**

Fences built in the wrong place can hasten erosion, invite lawsuits, allow livestock to escape, or simply not last.

In addition, boundary markers may have become lost or obliterated. The corners of existing boundary fences may be in the right place, but the line itself may run crooked as it goes through bushy, rolling terrain so you may be trespassing if your new fence joins up somewhere in the middle of the line. Others may have long-held rights of passage or use that they want to preserve. Even if they are subsequently proven wrong, it could make your fence-building a bitter experience. Parts of an original land parcel may have been conveyed out for construction easement (e.g., road widening) or rights-of-way (e.g., for buried utility lines), and all these need to be correctly determined.

There also may be the technical matter of where your property boundaries are in relation to natural features, such as waterways. Suppose for example that you are fortunate enough to have a flowing creek bordering your farm. There are very strict and complex laws defining exactly

. . . the mining regulations which were in force in the Klondike gold mining camp defined the most valuable class of claims—creek claims—as extending from "base to base of hill." As the country is of glacial formation, and what may have been at one time a sharply defined base of hill is generally overlaid with from 10 to 60 feet of slide matter, and as, moreover, the values involved were often enormous, the result was an endless succession of law suits on the interpretation of this particular definition. . . . I distinctly remember a well known member of the Dominion Geological Survey give evidence to the effect that, in his opinion, the "base of the hill" was half way up an adjacent mountain, basing his opinion on the probable position of said base of hill during the Pliocene Period.

Descriptions of Land: A Text-Book For Survey Students **by R.W. Cautley, D.L.S., B.C.L.S, A.L.S.**

where the land border falls. As with many areas of law, the results are not always what you might expect. In some jurisdictions, the boundary may be the "middle thread" of the stream. Where land borders any waterway navigable by commercial traffic (no matter if the last steamboat passed by one hundred years ago), the boundary might be the "ordinary high water mark." In other jurisdictions, including mine, the boundary is defined by the change in "character of permanent vegetation," so the actual boundary might be dozens of yards back from the water. The key point is that water boundary laws are very complex and change from state to state or county to county.

Another case of natural features as land boundaries occurs when it has been represented to you that your land begins or ends at the foot of a hills or range of hills, as in this example.

There can be complex legal decisions involved and it does not work to simply fence where you think it looks proper.

It may be reasonably thought that land descriptions and property marking has advanced since the days of the Klondike Gold Rush, and in built-up areas it undoubtedly has. But when you move to a new property way out in the country, you could be surprised by what you find slumbering out there. Some rural areas were surveyed and deeded at or before Klondike times and have been left unnoticed and undisturbed until things get stirred up by building new fence.

For example, according to information on the Minnesota Society of Professional Surveyors website (www.mnsurveyor.com), boundary problems often crop up regarding lakeshore property in Minnesota. This now very desirable lakeshore land was originally divided and sold one hundred years ago, coincidentally right around the time of the Klondike Gold Rush. In Minnesota as well as the Yukon, modern laws and regulations were not yet in place. The farther east and south you go, the older and chancier the surveys get.

It's not good enough to build fences according to what you thought were the property corners. All possible effort must be made to gather evidence of the original corner location. And for that, you really should have advice and a certified property plan from a surveyor legally qualified in your jurisdiction. The cost is small compared to the potential for ill will between neighbors, labor, time, and legal hassle involved with settling an issue over an incorrectly placed boundary fence.

Why this is relevant to you today is that:

The law provides that the original corners established during the process of the survey shall forever remain fixed in position, even disregarding technical errors which may have passed undetected before acceptance of the survey. The courts attach major importance to evidence relating to the original position of the corner, such evidence being given far greater weight than the record relating to bearings and lengths of lines. The corner monument is direct evidence of the position of the corner.

United States Bureau of Land Management Manual of Surveying Instructions, 1973. www.blm.gov/cadastral/Manual/73man/id166.htm

FINDING YOUR PROPERTY CORNERS

In the best case, easily identifiable iron posts will define the corners of your property. The newer the subdivision of land was made, the better the chance this will be the case. But the older the survey, the better the odds that some other method may have been used. Reference to the original survey plat is crucial to determine the original surveyor's method of marking the corner.

The law of property allows for an interesting array of potential markers, such as broken glassware or crockery, a marked stone, a charred stake, a quart of charcoal, or pieces of metal. Anything not native to the original local soil that provides durable evidence of the original placement of the marker constitutes a suitable memorial. In some cases where the original wood or iron post has been removed, the land surveyor may carefully remove thin layers of subsoil to find the darker topsoil that fell into the cavity made by the original marker. The original position is that important in the law of real property.

Even when a marker seems evident to the casual observer, you should get expert confirmation that it's the right marker. Sometimes what appears to be a property corner marker is actually the marker for something else, such as a road widening or utility right of way. If the post could not be set at the correct place because a fence post, large stone, or some other obstacle is already occupying that position, a reference marker may have been set instead with the offset bearing and distance noted on the plan of survey. If you start your fence from the wrong post or from a reference marker, your fence will of course be incorrectly placed.

In many areas, the original survey marker was either defined or referenced by an arrangement of pits and a mound of soil removed from those pits. The idea was, and still is, that even if the post is obliterated, the character of the soils in the pits and mound can provide useful evidence about the original position of the post. You might logically assume that the top of the mound would be the property corner, but that is usually not the case. Once again, the services of a surveyor qualified to practice in your jurisdiction are necessary for legally reliable location of your property boundaries.

Don't skimp when it comes to locating the original boundaries of your property. The cost of hiring a surveyor is small compared to the stress, time, and money you'll spend if you put your fence in the wrong place.

PROTECTING YOUR SURVEY MARKERS

Since replacing a property corner marker you've inadvertently destroyed involves a substantial cost for a new survey, it's logically worthwhile to protect any existing survey markers. Pulling the marker out and setting a fence post on the same spot is not the best way of proceeding because you've effectively destroyed the marker and could incur costs to get it reestablished. Your word on where the corner was is good evidence, but it may no longer be conclusive evidence. If over the years you forget or die without telling anyone, then some of the best evidence is lost.

The value in survey markers is also lost if you pull them out and stick them back in after fence construction is complete (and you could technically be fined or jailed for doing so). Markers are only legally valid if their position has been reestablished by a surveyor professionally qualified in your jurisdiction. If you feel you absolutely must have the fence corner exactly on the property corner, call a surveyor first. They can establish offset markers that will, once you're done fencing, allow the original corner to be unambiguously located and recorded.

A much easier way to go is to slightly offset the fence post to miss the survey marker. The sensible purpose of the very slight jog in the fence can be made clearly evident to any viewer by placing a brightly painted metal guard post near the survey corner. In addition, having fence corners slightly rounded also protects livestock from being crowded into a tight corner, either by accident (e.g., cattle in a storm or sheep spooked by predator) or bullying by a dominant animal (a noted problem with horses).

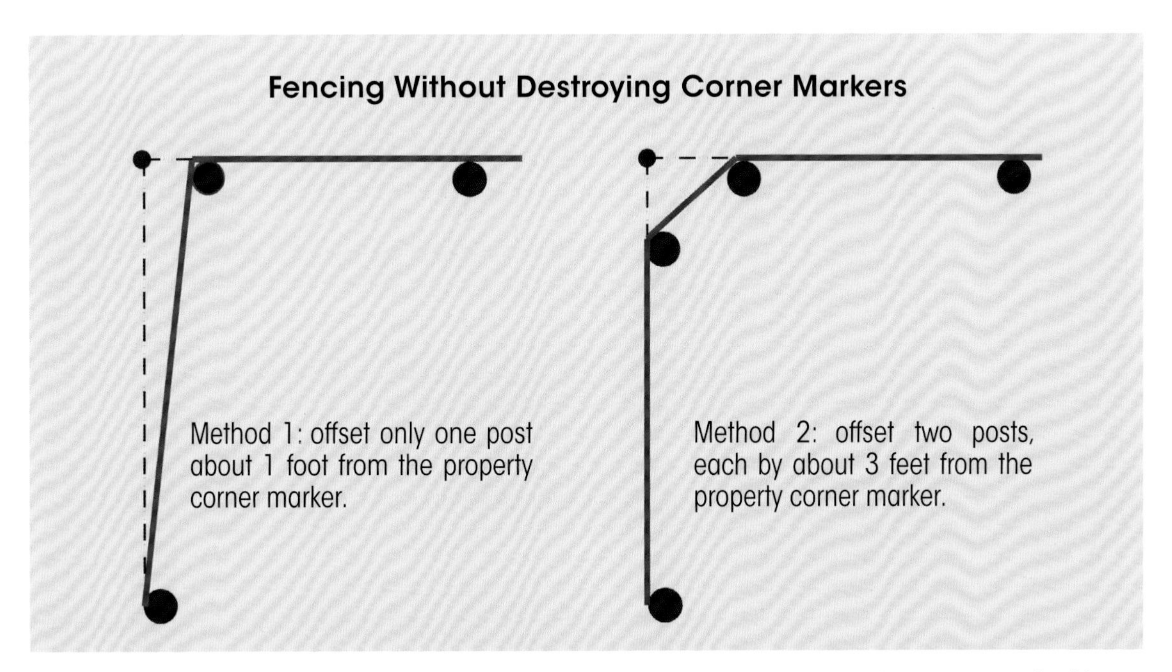

A small offset in the fence line achieves a lot in terms of preserving markers and avoiding arguments. In either style of offset, the amount of lost land is minimal—about 1/10,000th of an acre.

CHAPTER

2

PLANNING THE FENCE

No one kind of fence is best for all situations. Make a plan in order to select the best kind for the job. This grass strip fenced by high-tensile wire directs cattle to a pasture with a permanent barbed-wire fence.

To make sure the fence does the job you expect at a reasonable cost, it's essential to do a little planning first. The first point to settle is whether the fence is to be permanent, semi-permanent (to be moved or removed within a few years), or temporary, because this affects your choice of posts and fencing materials.

If the fence is to be temporary, such as a snow fence or the electric paddock fencing used in rotational grazing systems, then it is best to use materials that are easy to put up and take down. Electric fencing is ideal for this purpose, while barbed wire is quite difficult and time consuming to take down and roll up. Line posts can be the simple step-in types that are driven in with foot pressure. Corner posts need to be a bit stronger to stand the extra loads, but step-in posts, steel T-posts, or small-diameter wooden posts (pencil posts) are usually sufficient.

Semi-permanent fences might be required for boundary fence on rented land or to surround new trees for a few years until they are fully established. The materials still need to be easy to put up and take down, but the posts will need to be sturdy to bear the load for a longer time.

A permanent fence will require stronger posts treated to resist rot and set as firmly as possible in the ground. A permanent fence will need large posts and diagonal bracing at the corners to resist bending or being pulled out of the ground over time. The full range of fencing materials is available to consider: multi-strand wire, mesh, boards, panels, or a combination of several elements.

PURPOSE OF THE FENCE

The intended purpose of the fence will also have a great deal to do with your design and choice of materials. For example, a three- or four-wire barbed-wire fence that is fine for managing the movements of cattle is too dangerous for horses; poses no barrier at all to dogs, coyotes, or foxes; and is not

If you can get your neighbor to agree to share the cost of a boundary fence, you might be able to afford one of better quality.

that attractive in terms of beautifying the farmyard. Clearly, the purpose of the fence needs to be set down as part of the plan.

The purpose of the fence essentially comes down to listing what needs to be kept in and what needs to be kept out, and how the fence should look while doing so. But one more element can profitably be considered first: who and what is on the other side. You may be able to secure an agreement with a neighbor to share the cost of the fence. Since you're potentially bearing only half the cost, you may be able to afford a more advanced design using higher-quality materials.

Statutes vary from community to community on the responsibilities of adjoining landowners regarding a common fence. Some areas require boundary fences and some don't; some require adjoining owners to share costs of a fence while some don't. Equip yourself with locally relevant facts before entering into any discussion about shared costs of fencing. A call to your county office and a search through agricultural extension documents related to fences are an excellent starting point.

Even if you've determined costs are to be shared in principle, there may be proportioning to be done if one side wants a more or less expensive design of fence. You may want a horse-safe and beautiful but expensive white board fence, but the neighbor only needs and is willing to pay for a basic three-strand barbed-wire fence, or vice versa. Expect some negotiations with your neighbor.

Itchy animals rubbing against a fence do a lot of damage.

Design decisions at the planning stage can make the fence more visible to animals and improve strength and aesthetics.

Along with who is on the other side, there is the matter of what is on the other side. The main issue is which side of the posts the wire is to be placed on. Cattle, sheep, and other animals love to scratch themselves on anything, and the side of a fence post without barbed or electrified wire will do nicely. If allowed, the animals will gradually push the post sideways or snap it off entirely. Do not underestimate the destructive power of livestock, especially cattle. Just through rubbing, they can eventually push over dead trees of quite a large diameter.

Alternating the wires from one side to another as you go from post to post spreads the problem to both sides and is not that useful of a solution. Stapling on vertical strips of barbed wire is also very inconvenient and rarely used. If you find that livestock are going to be penned on both sides of the fence, an electrified strand on the side without wire or boards may be the best way to discourage destructive rubbing and scratching by those on the other side.

WHAT'S BEING KEPT IN?

Before going ahead with construction of any fence meant to contain animals, you need to learn all you can about the habits of those animals. It's not just an issue of them escaping, as inconvenient as that is. One of the saddest and most sickening things about an insufficiently planned fence is when an animal gets partway through or over but is then trapped in the fence or hung up on top of it, causing starvation or slow death from injury, such as the wire tearing into its belly while it's hung up on top of the wire. If your animals are pastured near a road, escapes could lead to vehicle-animal collisions that lead to expensive repairs to vehicles and potentially fatal injuries to both animals and humans.

One of the first principles of successful stock containment is to reduce the urge to get out. If there's good feed, water, and shelter inside the enclosure, animals are less likely to covet things on the other side of the fence and test the strength of the fence to get at them. This is one of the key principles in fencing for hard-to-hold animals, such as bison. Enhancing the benefits of being inside the fence won't be totally effective for animals that are naturally curious or have a desire to wander, but it's a good starting point. Animals that remain very persistent about breaking down fences, no matter how good they have it inside, may have to be culled from the herd.

Consult livestock management information, fencing supplier recommendations, and neighbors' experience to learn about fence construction details needed to successfully manage the animals in your care. State and provincial agricultural extension offices have plenty of good information, most of which is available for free on the Internet or at a nominal cost for printed copies. Enthusiast organizations, such as horse clubs or sheep producers' associations, also publish plenty of useful information that's kept current with members' experiences.

Any animals that are to be contained within the fence will have some peculiarities that have to be considered when you're planning your fence. For example:

Dogs can become very good at climbing fences. Additionally, any terrier breed will make a determined and usually successful effort at digging underneath a non-electrified fence.

Sheep can easily get their heads caught in wire mesh and aren't very smart about reversing to get back out. During weaning, sheep can become so overwrought that they crash through or get tangled up in fences.

Old-school wooden braces are strong, but best used only for cattle fence. Small, sure-footed animals like goats or coyotes can use the wide, flat brace to climb to the top of the fence and jump over.

Goats, especially billy goats, are expert climbers, which is something perhaps left in their genes from wild mountain-goat ancestors.

Pigs will accomplish remarkable feats of excavation with their snouts, including rooting right under the fence. If they are truly determined to get somewhere, pigs may also work themselves up into a "banzai charge" to crash through a fence. They are apparently willing to endure short-term pain in order to achieve the benefits they sense on the other side.

Horses may be contained by barbed or high-tensile smooth wire, but either kind of wire can be very dangerous to them. They are not that good at seeing thin wires, and may run into wires unless visibility aids such as flagging or wide tapes have been added. Colts are also very skilled at rolling underneath a fence whose wires do not extend close enough to the ground.

Cattle not only rub and push on the fence as mentioned above, but calves like to squeeze under, sometimes causing enough consternation to the cows that they will attempt to jump the fence to be with their calf. Cows are terrible at actually jumping a fence, and tend to get entangled in it.

Bison are not affected by barbed wire and will jump or knock down fences if they wish to get somewhere, but they may be trained to stay in an area.

Exotic stock, such as deer or elk, are such good jumpers that fence height becomes a critical planning issue.

With any animal, the opportunity for breeding may bring out unexpected abilities to get under, over, or through a fence that would ordinarily contain them.

Note that if you're planning to fence in mixed herds (e.g., cattle, sheep, llamas, and horses) or to rotate various kinds of stock on the area over time, the fence will have to be able to hold up against all escape problems presented, not just one. The rule of thumb is that if you build for the worst problem (e.g., goats), the other containment problems will be handled as well.

WHAT'S BEING KEPT OUT?

Along with keeping certain things in, fences often need to keep other things out. Back when farms were first being established in North America, the function of many fences were not so much to keep stock in as they were used to keep legally free-ranging cattle and hogs out of crops and gardens. The problem of exclusion was then

Yorkshire Sheep Roll to Freedom

Marsden, United Kingdom—Sheep on the fenced Yorkshire commons have learned to roll across 8-foot-wide cattle guards to get at gardens and other grazing areas, including the village park, bowling green, cricket field, and graveyard.

An eyewitness reports that the sheep lie down and just roll over and over the grids until they are clear. She added that once they have gotten free, sheep are not frightened and are quite unwilling to be driven away, even if they see a dog.

A National Sheep Association spokeswoman said: "Sheep are quite intelligent creatures and have more brainpower than people are willing to give them credit for."

From "Crafty sheep conquer cattle grids," BBC News, 30 July 2004 www.news.bbc.co.uk/1/hi/uk/3938591.stm

much worse because the law considered land to be open range, where animals could be pastured anywhere they felt like wandering. Under open-range law, it is the responsibility of landowners to fence in anything they did not want rooted up, trampled, or eaten. It has taken a long time to convert to the present closed-range type of law, under which owners have to keep stock confined. Today there are counties in a few states that still hold to the old open-range law.

Even in closed-range jurisdictions, the problem of keeping free-ranging animals out is still with us today in such forms as keeping foxes out of chicken yards, coyotes out of sheep pastures, low-grade stallions away from pedigreed mares, unwanted recreational users off no-till fields, and many more. Even some insects can be fenced out: steep-sided, plastic-lined ditches are one method for integrated control of Colorado potato beetles in potato crops.

One of today's most vexing exclusion problems arises in areas with many small holdings where the owners go to off-farm jobs and leave their dogs to run free. These normally well-behaved, lovable dogs often form a pack with other bored dogs, and the pack behavior quickly reverts to something resembling the dogs' wild ancestors. As difficult as it might be for the pet's absentee owner to believe, mass kills of chickens, mutilation of sheep, and running to death of cattle can be the result. It's all very difficult to deal with when the pack members, knowing their owners always come home at a certain time, run back home to wait innocently on the step. Some very bad feelings have been seen to erupt between neighbors over this problem, so dog-proof fencing may be an element to consider in your plan.

Fortunately, there seems to be a constantly improving solution to most types of animal predators: some type of electric fence in strands, netting, or a combination of the two. An electrified strand helps prevent close approach to the physical barrier.

Dogs are intelligent enough to quickly learn to avoid crossing electric fences.

Deer fence designs take advantage of the animals' poor depth perception to provide the illusion of a fence too wide to jump. In addition, attractants like special scents or peanut butter can be put on the electrified wire to train deer that they will get a shock when they touch the wire.

Moose often break down or jump over normal fences. However, research in Newfoundland has found that certain types of electrified rope are very successful at keeping them out of fields of cabbage, a favorite snack for moose.

Foxes can be excluded from chicken pastures with electrified mesh. The shocks they get from touching the wire often keep foxes from approaching the fence near enough to use their exceptional tunneling skills.

Raccoons can be kept out of sweet corn and other vegetables with a nose-height electrified tape or mesh.

Skunks, rabbits, and other small pests have less soil contact due to their lighter weight, but higher voltages and closer-spaced mesh are usually effective in keeping them out.

Electric fences are mainly a pain barrier, so truly determined or extremely hungry animals may blast through the pain to get through the fence. In that case, the fence needs to also provide a physical barrier, such as closely spaced wires, tight mesh, and/or a tall height. Electrified mesh is available as wide as 68 inches to make a fence tall enough to repel more determined deer, coyotes, wolves, mountain lions, bears, bobcats, raccoons, billy goats, and dog packs.

PLANNING ELECTRIC FENCE POWER

If you're planning to use an electrified fence, choosing an effective energizer is a key step. Modern energizer designs have become much better at maintaining effective voltage than earlier types, so don't base your plan on buying a used older energizer, no matter how attractive the price is at an auction.

Manufacturers rate the power capacity of fence energizers by several methods, such as miles of fence powered, voltage output, joules, and effective voltage under varying resistance levels. The first three measures are useful in selecting the right size unit within one manufacturer's line of energizers, but they are not quite as useful for comparing among brands. Ideally, energizers would be rated based on the last-mentioned criterion of actual fence-line voltage measured under several known levels of resistance and also be rated in comparison to one other. But this information is not currently available, and given the high cost of unbiased third-party testing, it's not likely to become available any time soon. The best option at this point is to choose a reputable energizer manufacturer and follow their recommendations on what size to use.

A saying from New Zealand, a world leader in development and use of high-tensile electric fence designs, is that North Americans typically "overbuild and underpower" high-tensile electric fence. That is, we tend to use too close of a spacing between line posts so the fence is less able to flex under strain, and then skimp on energizer power because the fence appears stronger. Take a tip from the Kiwis and go with a bigger energizer and fewer posts.

It's a little more straightforward to make plans for what power source to use for the energizer. Where the energizer can be located close to 110- or 220-volt power, a plug-in energizer is usually the best choice. They are cheaper per unit of output power than battery units, and reliable mainline power means the fence charge is fairly reliable.

There are many good choices now in electric fence energizers. Before you buy one, make a detailed plan: you don't want to have to upgrade a few months later. If you are deciding between two similar sizes, go with the energizer that provides more power.

If the fence is too far away from a plug-in, a battery-powered energizer becomes part of the plan. Very small units with less than 1 joule output may be powered by dry cell batteries, while larger units up to about 9 joule output can be powered by 12-volt automotive type or deep-cycle RV batteries.

When planning to use a 12-volt battery-powered system, a solar panel helps keep the battery charged in place. Without a solar charger panel, large battery-powered energizers can fully discharge a deep-cycle 12-volt battery in less than a week. To avoid having to bring out a freshly charged battery, hook it up, and take

Innovations in electric fencing include items such as this solar panel, which is used for charging electric fencing.

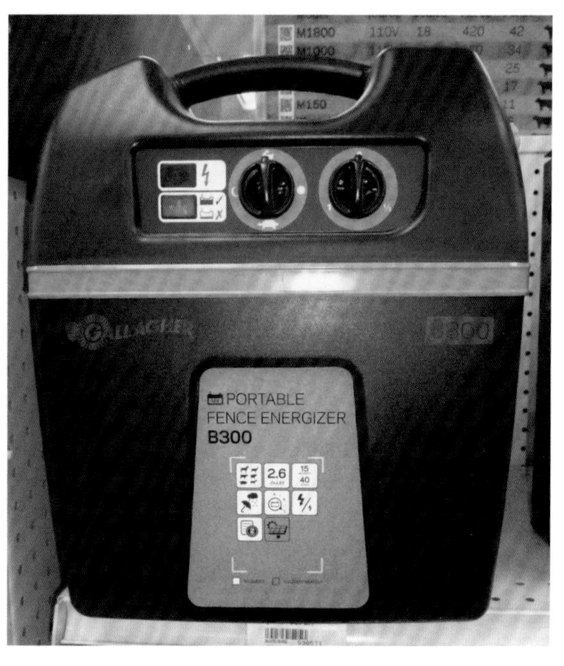

Modern portable energizers are useful for electric fences and when changing paddocks (such as when using rotational grazing methods).

the discharged one back to the shop for recharging, consider the solar panel option.

A general recommendation is that for each joule of energizer output, allow 7 watts of solar panel capacity in high-sunlight areas or 10 watts in low-sunshine conditions. Even in a generally high-sunlight area, the 7-watt-per-joule panel may not be enough to keep the energizer power source fully charged during extended periods of cloudy weather, so the larger-size panel may be needed in your plan.

Getting the full effectiveness out of your energizer depends on planning out a properly sized grounding system as well. Your plan should provide for a minimum of 3 feet of ground rod per joule of energizer output capacity. Ground rods are typically ½ to ⅝ inch in diameter, 6 to 8 feet long, and placed a minimum of 10 feet apart. Place them where they can be in contact with moist enough soil to support plant growth all year round because moist soil conducts electricity much better than dry soil. Placement on the north side of a shed under the drip line is an ideal location. Plan so that the energizer ground rods are also at least 25 feet from the nearest power line ground rod.

If your soils are so hard or rocky that it would be very difficult to get the ground rods more than a few feet in the ground, your plan can be modified to have the ground rods lying in shallow trenches where they can be in contact with moist soil year round. As before, placing them on the north side of a shed under the drip line works well.

Also plan on installing lightning protection for the energizer. Both battery-powered and plug-in energizers need power

With lightweight electric mesh, gates are only needed for animal movement. People can easily cross anywhere by first stepping on the middle of the mesh to pull it partway down.You can then step on top of the mesh to get it completely out of the way. Just don't try this with bare feet.

spike protection on the output side to prevent lightning voltage from striking a fence and traveling back into the energizer. Plug-in energizers also need spike protection on the input side to prevent a lightning strike on the power lines from traveling into the energizer. The same kind of surge protector used to protect computers can be used to protect the energizer form lightning, as well as the variations in voltage flow that are common in rural electric systems. These smaller surges are almost as hazardous to electric fence energizers (and computers) as lightning strikes.

For lightning protection on the output side, try to place the energizer somewhere that is generally at lower risk of lightning strikes, such as low-lying ground or in a shed protected by lightning rods. An induction coil or lightning choke should be installed near the energizer in the wire leading to the fence. Either of these devices functions as an inline resistor to the lightning surge and can reverse the energy flow back to the fence line and away from the energizer.

PREVENTION OF DIGGING

For animals that burrow, such as foxes, dogs, or pigs, aprons of mesh can be laid on the ground outside the fence to prevent digging. Research in Australia has found that horizontal buried aprons are more effective than those that are buried vertically, such as in a trench. With vertical buried aprons, excluded animals sometimes continue to burrow down until they are able to pass under the apron. With the horizontal apron covered with only a thin layer of earth, animals quickly hit a dig-proof barrier even if they try in several places.

Horizontal aprons also involve a lot less work in construction because there is not a trench to dig. Aprons of strong metal mesh can be secured to the ground surface by pegs, rocks, a covering of soil, or by letting grass grow through the netting. Where soil is already hard for animals to dig, securely pegging the apron to the soil surface is usually sufficient. But where soil is soft and easy to dig, an apron that is buried just below the surface has been observed to be better because it prevents animals from easily locating the edge of the apron or gaps in the wire. Buried aprons do corrode faster than surface-laid aprons, so mesh with a thicker layer of galvanized coating may need to be part of the plan.

The anti-burrowing apron can either be a strip of mesh completely separate from the fence or wide enough that an unburied portion can be bent upward to be attached to the vertical part of the fence. The second method is stronger, but the disadvantage is that when the apron corrodes, a larger section of mesh will need to be replaced. Attached or not, when the apron corrodes, do not lay a new apron on top of the old corroded layer. This will make the new netting corrode considerably quicker than it would otherwise.

(Above) A bar across the top of the entrance gate adds to the good looks of this entrance and also adds strength that helps keep gate posts from leaning inward. Just make sure the entrance height is enough to allow tall equipment to pass through. *(Below)* If a gate isn't used very often, you could tie it shut with rope. But this setup is slow to operate and ropes tend to deteriorate in sunlight, which leads to a gate that might fall open at any time.

GATES

Access gates often prove to be weak points in exclusion fences, so the gate needs to be as tightly designed as the fence itself. Gaps in the gate bars or mesh, below the gate, or between the gate and its supporting post must be no less than the gap sizes in the fence. If that proves to be difficult, corrugated sheet metal can be hung on the fence to deter penetration or climbing. When vehicle tires form ruts beneath the gates, a solid wood or concrete pavement should be laid down to eliminate gaps. This has the added benefit of preventing a muddy trench from forming during periods of rain.

Gates that are supposed to permit the access of people or vehicles can be problematic if you rely on people to

A more durable home-built gate latch consists of lengths of metal pipe and chain welded together. The gate is pulled partially up by hand and then the end post can be fitted to the place where the pipe and chain meet. The pipe is then used as a lever to pull the fence fully tight and is secured by the remaining free end of the chain.

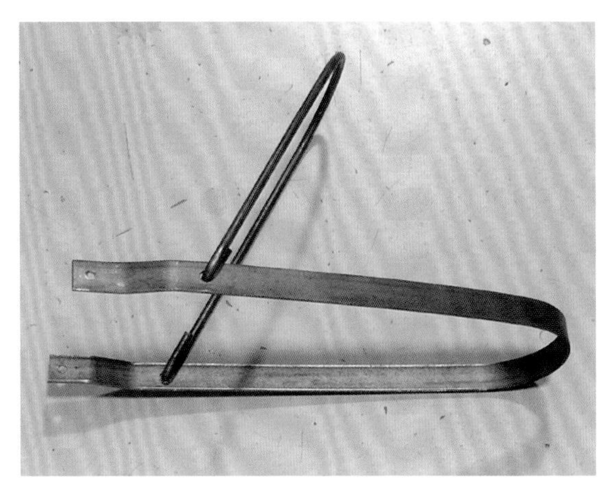

Commercially made gate latches are available for about $10 and are very easy to install and use.

The leverage provided by the large handle makes it easy to pull the gate until it is tight. A bent pivoting nail on the horizontal wooden rail latches the handle in place.

In use, the large handle is lifted up until the hoop can be fitted over the gate post. A protruding staple on the post keeps the hoop from accidentally slipping upward.

This fence has been installed with an excessive bottom gap to allow it to swing toward the upslope. That's all right for containing large cattle, but it leaves too big a gap for calves or smaller livestock. If the gate swung down slope, the gap could be made smaller.

adequately close the gate. Pedestrian gates should be spring-tensioned in order to close and latch automatically. The grate-type cattle guard or "Texas gate" (as it's known mainly outside Texas, strangely enough) is more effective for controlling the movement of large livestock in than it is for excluding other animals because it takes advantage of the reluctance of cattle to walk over an area of alternating light and dark. Other animals that don't have this sensory issue won't be much deterred by the grate.

SOILS, MATERIALS, AND MACHINERY

The plan for an effective fence is not of much use if you can't afford to build it, so your plan will also have to take into account material and cost factors. One of the first items to review is whether you can get the posts into the ground and rely on them staying there. A low-budget fence project may stall after the exhaustion of digging the first hole with a shovel and bar or pounding in the first post using muscle power.

A few small test holes will provide a general idea of whether the ground is hard or stony, but be aware that soil conditions can change within a few feet, especially if you're going up or down a slope. A more reliable survey of soil conditions can be made by inspecting whether a neighbor's fence post was dug in or pounded in. Mounds near the post are good indicator that they were dug and tamped in, while compressed wood fibers on the tops of posts are evidence that posts were pounded in. If possible, ask neighbors if they had any difficulty with hard, loose, or stony soils. It's a peculiarity of the countryside that some neighbors know as much about the condition and history of your land as the seller did–sometimes even more.

The soft, wetter soils in boggy areas pose other problems. The soil doesn't grip the posts well enough to prevent shifting, frost tends to heave the posts up out of the ground, and posts rot faster in wet soil. Posts need to be longer and larger in diameter so they can be pounded in far enough to reach solid ground, and you can count on more frequent replacement. A better solution may be to use electrified fencing, which does not require such strong posts.

Time is another major factor in the practicality of your fencing plan. With so many things to keep you busy on a small farm, calling in a contractor might be a very good idea, even for part of the job,

(Above) Special planning is needed wherever a fence crosses low-lying ground. If special care isn't taken with post installation, the posts can pull out of boggy ground and make the fence lean over, as in this example. *(Below)* Even if the ground isn't wet right now, consider drainage patterns that could lead to flooding in the area of the fence.

such as pounding posts or power-augering a long line of post holes. A neighbor who already owns the necessary equipment may be willing to take on the contract for ready cash. Installing posts represents a big part of the time on a fencing project, while installing wires goes much quicker. Using the services of a fence post contractor gets one of the hardest parts out of the way faster and can free you from having to buy specialized machinery, such as a post pounder or post-hole auger. It also completely eliminates your having to operate these machines, which both have some pretty major safety concerns for an untrained and/or inexperienced operator. For more information on how to use a post-hole auger and post pounder, see *How To Use Implements on Your Small-Scale Farm* (Motorbooks, 2005) and *The Farm Safety Handbook* (Voyageur Press, 2006).

If you can afford to wait, another option is to grow a fence. Using thick, thorny hedges to fence fields was a common practice in parts of Europe and the British Isles and still is in some areas. Immigrants brought this technology to North America, where it was employed extensively, especially in the less-forested areas where rail fences were expensive to build. Before the invention of barbed wire, many miles of tough, thorny Osage orange hedge were established in the Great Plains states. Young trees were planted close together, then aggressively pruned to promote bushy growth. The eventual result was a fence that was "horse high, bull strong, and hog tight," which meant it was tall enough that a horse would not jump it, strong enough that a bull would not push through, and so dense with thorns that a hog could not squeeze through. These living hedges were also quite attractive as a landscaping element and served as an important wildlife habitat.

The popularization of barbed wire in the 1880s made hedge fences (along

In permanently marshy ground, an electric fence may be a more suitable choice because the lightweight wires exert much less strain on posts.

Besides acting as a fence, hedgerows can create a beneficial microclimate for crops nearby.

with most other types) obsolete, but not everyone liked using or seeing barbed wire. As late as 1939, the USDA reported that approximately 39,400 miles of hedge fence graced the Kansas landscape. It took the advent of large farms in the 1950s to really make hedge fencing decline. Hedges make it inconvenient to maneuver large tillage equipment, and the powerful engines and front-end loaders on the new, larger tractors made it easier to remove the hedge.

But the technology of a living fence is still available for those who want to use it. In a fortunate blend of old and new methods, the modern technology of semi-permanent fencing can be added to provide instant fencing until the living fence gets fully established.

ENVIRONMENTAL ISSUES

Environmental issues may also affect certain elements of your plan of how to build the fence. For example, natural-gas extraction and processing activity in my area has increased atmospheric sulfuric acid content to the point where ordinary steel barbed wire corrodes much more quickly than it did in the past. To prolong the lifespan of wire fences, wire with thicker galvanizing coating is needed.

Other environmental problems to be considered may be local problems with flash floods in gullies or what post materials last the longest. Consult with neighbors and local farm supply outlets to determine any special concerns in your area.

(Above) The classic white-board fence looks good but requires a lot of maintenance. *(Below)* Even a utilitarian barbed-wire fence can be made more attractive with various decorations, such as these farm hats. Old boots nailed to the tops of fence posts are another popular decorating possibility.

AESTHETICS

Everyone has their own idea of how their farm should look, and if your fence is in a prominent place, it may play a prominent part in the appearance of your place. There's a definite difference in the "statement" made by crooked posts made from dead trees versus a straight row of stout posts or saggy, rusty wire versus white-painted boards.

Make sure that your plan reflects your personal style and can be reasonably expected to do so after several years exposure to animals and the weather. Even the simplest fence involves a lot of labor and time to redo.

Fence posts also provide a place to improve habitat for native birds, which are not only an aesthetic improvement but often consume large numbers of insect pests.

TEN ESSENTIAL PLANNING POINTS

Use the answers to these questions as a starting point for making your fence plan and to narrow down the wide list of options available at your fence materials supply outlet.

1. Is the fence to be permanent or temporary? If it is to be permanent, how many years do you expect it to last without complete replacement?
2. For boundary fences, are the property corners clearly marked?
3. Are there any adjacent property owners to be consulted with about part payment or the fence design?
4. Are there any legal requirements or community concerns affecting the type of fence being put up?
5. What needs to be kept in and how do they behave toward fences? For example, will the animals get their heads stuck in wire mesh?
6. What needs to be kept out (e.g., other stock, roaming dogs, foxes, recreational users) and what kind of a fence is needed to keep them out?
7. What kind of soil are you dealing with? Can you easily install all the posts necessary, or would hiring a contractor with specialized machinery be faster and safer?
8. Do you need the fence right now or can you wait for a "living fence" to grow?
9. Are there any local climate issues to be concerned with?
10. You may be looking at this fence for a long time. Is it something that fits with how you want your farm to look?

ASSESSING EFFECTIVENESS

The military dictum that "no battle plan ever survives contact with the enemy" might equally well be applied to the planning of fences. As well-planned as a fence might be, its true effectiveness may not become apparent for some time after its contact with "opposing forces," in this case the animals being contained and/or excluded.

Animals used to roaming through now-enclosed land or who see a tasty meal on the other side of the fence have plenty of incentive to try new tactics if at first they do not succeed. Initially, most animals that encounter a fence will first attempt to crawl under it, squeeze through what looks like any gap, or push against the fence to knock it down. Jumping is a high-energy tactic with risk of injury on landing site, so it's usually employed only as a later tactic or in desperate situations.

Since a much less energy-intensive tactic for those being excluded is to test the lower parts of the fence first, the lower sections of the fence need particular attention in the plan. Wires may have to be spaced closer together or carry a stronger electrical charge. An additional "stand-off" wire arranged to be several inches to a foot outside the main fence makes approaching the main fence wires much more difficult.

Fences near trees are subject to damage from limbs falling during windstorms. The projective limbs also attract livestock to the weakened point of the fence because another convenient scratching post has suddenly been provided.

The effectiveness of electric fences can be compromised by contact with green vegetation, which provides an electricity path. Keep vegetation cleared away and, if necessary, use a soil-sterilizing chemical under the fence.

Fence corners are another area of high pressure because excluded animals often walk along the fence line until they reach a corner and attempt to cross. Old-school diagonal wooden corner braces can provide a convenient catwalk for animals to climb up to the top of the fence and then jump down to the other side. Corners made by fences meeting at an angle of 60 degrees or less let nimble predators, such as foxes, brace their feet against adjacent portions of fence so they can climb it. Research has also found that some individual animals learn how to get past fences by watching successful crossings by animals of the same species, so don't provide any opportunities for them to see how it's done.

When monitoring the fence after construction, pay close attention to existing animal trails that cross the fence. These are areas with high risk of intruders trying to gain access or livestock smashing down the fence to go where they are used to going to graze or drink. Also watch for signs that livestock are starting to cluster at certain areas. Look for heavily grazed or trampled areas. These are also likely escape areas either because animals get bullied in these areas or because the herd is there more often. These fences may need to be beefed up at these points.

CHAPTER

3

ASSESSING WHETHER TO REPAIR OR REPLACE FENCES

You may be able to renovate existing fences at reasonable cost if the posts are in good condition.

Tightening and splicing mesh may be all that's needed to bring a tumbledown fence back into efficient service.

Some parts of your property may already be fenced, and continuing to employ these existing fences can allow a considerable savings of time, money, and labor. But before turning any livestock into the enclosure or relying on the fence to keep other animals out, it's worth doing a detailed assessment of whether the fences are actually okay as-is, need a few repairs, or should be replaced altogether. You may find there are only a few spots that need repair or that the fence is really much weaker and more decrepit than it first appears. The worst way to find out is to turn livestock into the fenced area and then have to round them up when they bust through the weak parts. This leads to a lot of stress and potential injury for both you and the animals, including the issues of animals getting out onto roads and being struck by vehicles.

To properly assess the condition of the fence, there's no substitute for close-up post-by-post inspection. Having an ATV or horse to ride while you do this makes the work go a whole lot faster, but walking is just as good. As you go, keep these points in mind:

Does it look like good materials and construction methods were used? A badly built fence can quickly become a long stretch of more trouble than it's worth.

Look at each post individually, especially corner posts and their bracing. Is the post rotting, tipping over, or pushed higher out of the ground than others? If you push on it, does it move? If you can move it easily, heavy livestock or fast-running deer will find it even easier to knock down.

Are the wires or boards still firmly attached? If staples or nails have pulled partially out, test their condition by hammering a few back in, then pull on the wire or board to see if they still hold.

Are there stretches between posts where wires are sagging or badly corroded? Are the boards badly warped, cracked, or rotted?

Are there any areas of ground that show signs of animals often getting through? Look for signs like areas of ground that hooves have trampled bare or big gaps at the bottom of the fence.

Check for soundness at the ends of rigid fence boards, where rot and splits are most likely to develop.

(Above) If your board fence has a board that falls down, the gap can be temporarily patched with lengths of rope, as shown here. Eventually, however, the board should be reattached with the proper method. Animals may be able to sneak out if the electric current fails for a long time. *(Right)* Methods of attaching fencing boards are illustrated here. On the left is the standard polymer clip that houses the end of the board. On the right, the end of the board sits in a joist hanger, available at any building supplies center. Joist hangers and other similar sheet metal pieces for building are low in cost.

Once you have a reliable idea of the condition of the fence, you can estimate whether it's time to rip it up and replace it or to make a few key repairs. For purposes of comparison, a cost guide for new fence construction is provided on pages 42-43.

(Left) Minor breaks in electric mesh fences can often be easily repaired with the splicing material supplied with packages of new mesh. *(Below)* Crossing the existing wires is a quick and simple method of tightening wires for a temporary fix.

If a gate post is only leaning and not broken off, it may be possible to pull it plumb again by installing this type of turnbuckle assembly. Turnbuckles of various sizes are commonly available at hardware and farm supply stores.

Twist-in battens (also known as droppers) keep spacing intact in sagging barbed wire. Battens are commercially available at low cost in various lengths.

Vertical members in ordinary barbed-wire gates do not require a great deal of strength, since they only act as battens to keep the wires from getting tangled when the gate is opened or closed. Small-diameter pieces of local wood are effective and keep the gate light enough for easy operation.

Estimating Costs for Fence Installation and Repairs

Except where noted, the information provided here is based on "Fencing Materials For Livestock Systems" by Susan Wood Gay, extension engineer, and Rick D. Heidel, extension agent, Animal Science; Virginia Tech.

HOW TO USE THE CHART:

1. Refer to the type of fence you are considering: permanent or temporary, wire or electrified, and so on.
2. Find the "cost index figure" for that type. This indicates the cost relative to other types. For example, fence with a cost index of 25 costs about twice as much per foot as fence with an index of 12.

Note: "Expected fence life" is based on a combination of post and wire life expectancy in a humid climate. In dry climates, fence life will generally be longer because wires don't rust as badly and posts don't rot as quickly.

PERMANENT FENCES

Type: Barbed-wire fence, wire spacing between posts maintained by wire tension only
Expected life: 33 years
Maintenance requirement: High

Number of Strands	Gauge of Wire	Cost Index
3	12.5	12
4	12.5	13
5	12.5	14
3	14	19

Type: Barbed-wire fence, wire spacing between posts maintained by battens
Expected life: 33 years
Maintenance requirement: Medium

Number of Strands	Gauge of Wire	Cost Index
4	12.5	8
6	12.5	10

Type: Woven wire, 6 inch spacing
Light weight
Expected life: 19 years
Maintenance requirement: High

Stay Height	Cost Index
26 inches	14
32 inches	15

Medium weight
Expected life: 30 years
Maintenance requirement: Medium

Stay Height	Cost Index
26 inches	14
32 inches	15
39 inches	18
47 inches	20

Heavy weight
Expected life: 40 years
Maintenance requirement: Low

Stay Height	Cost Index
26 inches	19
32 inches	21
39 inches	23

Type: High-tensile smooth wire fence, 12.5 gauge wire
Expected life: 30 years
Maintenance requirement: Medium

Number of Strands	Cost index
3	4
4	5
5	6
8	10

Note: Cost index may be lowered even further by adopting wider spacing between posts, as per proven practices in New Zealand and Australia.

TEMPORARY FENCES

Type: Barbed-wire fence, 2- or 4-point, 12.5 gauge wire
Expected life: 30 years
Maintenance requirement: Medium

Number of Strands	Cost Index
1	4
2	5

Type: Poly ribbon, 7/8 inch, 6 wire
Expected life: 3 years
Maintenance requirement: Medium

Number of Ribbons	Cost index
1	3
2	6

Type: Poly wire (stainless steel wires)
Expected life: 5 years
Maintenance requirement: Medium

Number of Strands	Gauge of Wire	Cost Index
1	6	2
1	9	3
2	6	4
2	6	5

Rules for Estimating Fence Costs

Adapted from "Estimating Beef Cattle Fencing Costs" by Lance Brown, engineering technologist, British Columbia Ministry of Agriculture, Food, and Fisheries.

Materials for a barbed-wire fence cost about twelve times the cost of wire alone. For example, if wire is $0.04 per foot, materials will run about $0.48 per foot.

Installed cost is about three times the materials cost. For example, if fence materials cost $0.48 per foot, total installed cost will run about $1.44 per foot.

Materials for high-tensile smooth wire fence cost about eighteen times the cost of wire alone. For example, if wire is $0.02 per foot, materials will run about $0.36 per foot.

Installed cost is about three times materials cost. For example, if fence materials cost $0.36 per foot, total installed cost will run about $1.08 per foot.

Materials for electrified high-tensile smooth wire fence cost about seven and a half times the cost of wire alone. For example, if wire is $0.02 per foot, materials will run about $0.15 per foot.

Installed cost is about three times materials cost. For example, if fence materials cost $0.15 per foot, total installed cost will run about $0.45 per foot.

EVEN DO-IT-YOURSELF LABOR ISN'T FREE

When comparing the cost of a new fence to the cost of repairs, don't forget to include the cost of your own labor in doing those repairs. For example, if fence repairs would involve about twenty hours of your time, would you be better off devoting that resource to some other opportunity that's more profitable or enjoyable, like training horses or marketing farm products?

This is not just an excuse for avoiding hard work. Any farmer can tell you that there is never enough time in the day to do the many tasks required. You need to carefully allocate time to avoid becoming broke, stressed out, or both. If a relatively uninteresting task, such as digging post holes and splicing wire, consumes too much of your time, you can easily end up short on time for more critical and/or enjoyable tasks, such as planting, caring for livestock, or whatever it is that led you to choose farming as part of your way of life.

When you're assigning a cost to your own time, remember that you are much more than an unskilled laborer. What would it cost you to hire someone to do everything of which you are capable: caring for livestock, operating machinery, making crop and planting decisions, ordering and purchasing and supplies, keeping accounts, and so much more?

If you assume your time is only worth what you'd pay the cheapest labor you could, there's a very good chance you'll devote more time to doing the tasks that only require cheap, unskilled labor. The appearance and productivity of the farm will likely reflect that drift to the bottom.

A more sensible approach is in suggested farm extension publications, such as a University of Wisconsin paper relating to a value-adding operation on an integrated small farm. A figure of $18 per hour is used in this 1994 study.

Another useful idea of the true cost of "free" time is given by research on the dollar value of a volunteer hour: $18.04 for 2005.

TYPICAL LABOR REQUIREMENTS FOR FENCE INSTALLATION AND REPAIRS

Except where noted, the information in this chart is based on "Fencing Materials For Livestock Systems" by Susan Wood Gay, extension engineer, and Rick D. Heidel, extension agent, Animal Science; Virginia Tech.

Insert battens into existing wires by placing the top wire between the strands of the batten and then turning the batten to thread it down toward the next wire.

Set the next wire at the desired spacing and continue to turn the batten to automatically thread the wires in place.

Turn the batten until the top wire is in the top loop of the batten and all wires are engaged at the correct spacing. This batten is used for demonstration purposes and is actually a bit long for the fence. It can be clipped off with wire cutters.

Labor Requirements for Fencing

ITEM	TYPICAL LABOR REQUIRED
Line posts	
Machine drive new post	6 minutes
Hand tamp post	20 minutes
Pull out old post, enlarge hole for new post	30 minutes[1]
Dig new hole by hand	60 minutes[1]
Permanent brace posts	
Dig new hole by hand	60 minutes
Hand tamp post	20 minutes
Machine drive new post	8 minutes
Construct brace assembly	
Single span	45 minutes
Double span	1 hour 15 minutes
Woven wire	
	30 seconds per foot
	6 minutes per post
Barbed wire (per wire)	
Unwind and stretch	10 seconds per foot
Fasten	1 minute per post
High tensile wire (per wire)	
Unwind	1 second per foot
Stretch	10 minutes per stretch point
Fasten	1 minute per post (barrier fence)
	2 minutes per post (electric fence)
Electrified rope	250 feet per hour after fence for horsesposts are in place[2]
Electrified netting (poultry, wildlife)	6 minutes per 164 foot roll[2]
Electric fence controller, install	2 hours per unit

Temporary and Semi-Permanent Fences

Note: these figures include post installation, wire application, and stretching

Horse fence, electrified tape	400 feet per hour[2]
Horse fence, rope	200 feet per hour[2]
Wildlife fence (to exclude small animals)	250 feet per hour[2]

[1] Personal experience

[2] Premier Sheep Supplies Fencing 2006 Catalog (www.premier1supplies.com)

SPLICING BARBED WIRE

The first step in repairing a break is to splice a new piece of wire onto a broken strand. With the old and new wires pinched between the legs of the fencing pliers, wrap the new wire eight times around the old wire to make a secure splice.

Form a loop in the other end of the new wire and wrap the free end of the loop back around the wire to form a secure connection.

Insert the other end of the broken strand into the loop and pull it as tight as you can by hand.

Connect a tightener that will pull the wires to full tension (about 250 pounds of force).

Slide one end of the tightening device as far back along the wire as possible, then flip the spring-tensioned jaw down to grip the wire tightly. Move the tightener handle to move the ends of the wire closer together.

Once the wire has been completely pulled toward the loop, bend the broken strand back to form another loop. Grip the two wires of the loop between the jaws of the fencing pliers and wrap the strand around itself to make the final connection. Remove the tightening device.

The low-tensile wire used in barbed-wire fencing breaks quite often, especially under the pressure of animals scratching themselves against it or pushing on it to reach grass on the other side. Splicing barbed wire is a skill very necessary to keep barbed-wire fences operational for their full expected service life.

The photo sequence here shows the procedure. One additional caution is that the older and more corroded the wire gets (and the lower the quality of the materials when first installed), the more likely it is to snap off when bent to form loops. You may end up going back quite a ways from the break to find wire that will satisfactorily bend. For inserting lengths of new wire between loops, carry a length of relatively new, good-quality wire so you're working with flexible, easy-to-splice wire.

If you become really good at splicing barbed wire, you might consider trying for a world championship at the Barbed Wire Splicing Contest in LaCrosse, Kansas. This event is part of the annual Antique & Barbed Wire Swap & Sell at the Kansas Barbed Wire Museum. Contestants demonstrate their strength and speed in splicing a simulated barbed wire fence with only leather gloves as their tools. Beauty doesn't matter in this contest, because it doesn't matter how the splice looks: it just has to support a 70-pound weight suspended from its center. For information, contact the Kansas Barbed Wire Museum, 120 W. First Street, LaCrosse, KS 67548; 785-222-9900; www.rushcounty.org/BarbedWireMuseum/index.html.

If the wire to be tightened is near a post, it can be pulled by gripping it between the legs of the fencing pliers and using the pliers as a lever against the post.

A crowbar can also be used as a lever to apply tightening force. The nail-puller notch in the crowbar will grip the twisted wire.

An older but still effective tightening device is this block-and-tackle arrangement that uses pulleys to multiply the pulling force applied to the rope.

The wire gripper at the free end of the block-and-tackle grips the wire more tightly as more pulling force is applied.

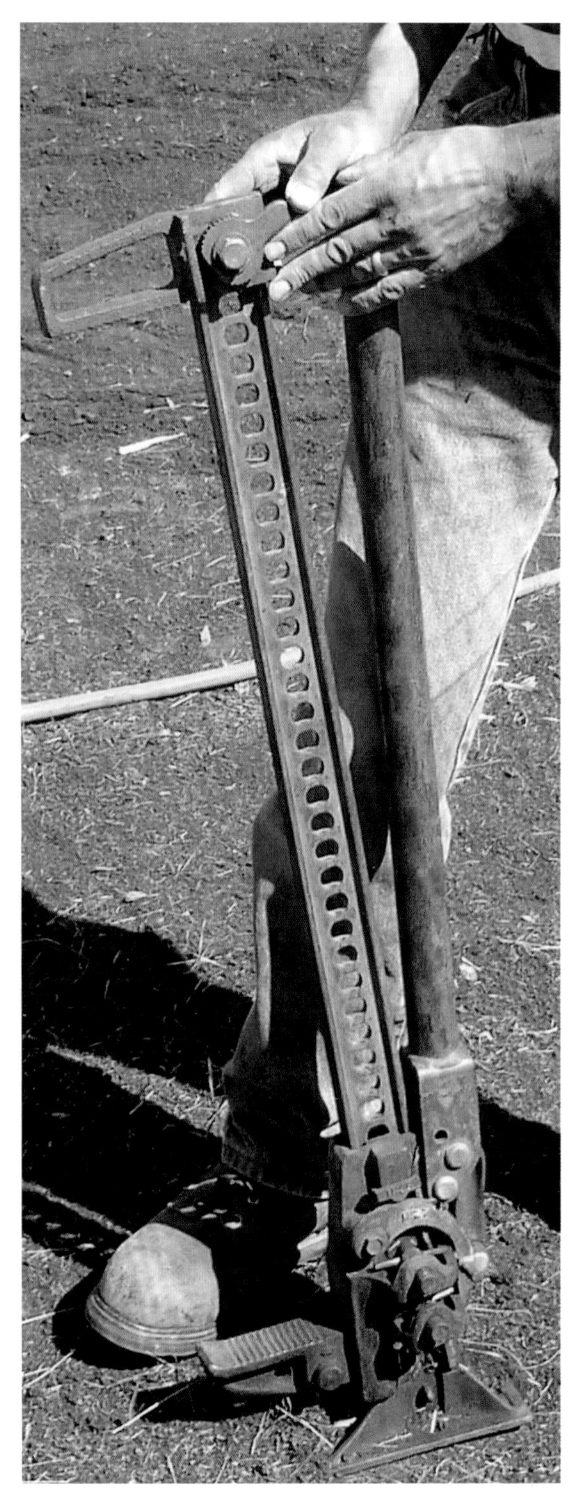

Some mechanical farm jacks have a wire-gripping device at the top of the jack so that they can also be used as wire pullers. The weight of this type of jack does make them awkward to use in a horizontal position as needed to tighten wire.

Combination barbed-wire/mesh fence will require separate tightening techniques for the two fencing elements. Start tightening at the top and work your way downwards to minimize loosening wires you've already tightened.If an older post is loose or broken, you can simply place a new post near it and leave the old one standing.

If the brace wire is not too corroded, you may be able to turn it a few more times to tighten it up again. Severely aged wires may break if twisted very much.

A lack of bracing has caused this corner post to be pulled sideways due to tension from the fence wires. Bracing by lashing to a nearby utility pole is not recommended, as it will bring complaints from the utility company.

(Above) This is a brace failure due to the corner post being too small and/or not set sufficiently in the ground. A new corner post and proper brace installation may be needed for successful fence renovation. *(Right)* Setting the end of a wooden brace too high, as in this example, may keep the end of the brace board drier and less prone to rotting, but it can also lead to weakening of the post that the brace board pushes against.

1

Brace assembly failure due to a loose post at the middle and right has led to difficult operation of the gate (visible folded back behind the posts).

2

Tearing down the brace assembly reveals that the middle post has snapped completely off at ground level. Temporary bracing of the middle post has been made with a steel T-post.

3

The two offending posts have been removed and replaced with strong, rot-resistant surplus railway ties that have been firmly tamped with gravel into expanded postholes.

4

Pressure-treated landscape timbers are notched into the new posts and secured with corrosion-proof deck screws to form top bars for the renovated brace assembly.

5

Twisted diagonal wires are installed to finish the brace. In many cases, the lever used to twist the wire ("twitch stick") is left braced against the top rail, but in this case, it was not necessary. The high-tensile fence wires are tightened with the ratchets visible between the two new end posts.

The gate hanger pivots are well worn but still in good enough condition to be used again. They can screw into new pilot holes drilled in the end post.

The gate is lifted up and dropped onto the hanger pivots. The pivots can be screwed in or out to adjust the lower gap on the gate.

A New Zealand–style (Kiwi) gate latch like this will be installed to provide secure latching and easy one-handed opening.

When the center link of the Kiwi latch is lifted, the curved latch piece is easily lifted free. When the latch piece is dropped back in, the center link automatically drops into place to lock the latch.

CHAPTER

4

TYPES OF FENCE

A review of the many types of fencing available helps keep the costs and labor in line with your goals.

Once you've gone through Chapter 2 to flesh out the details of the fence you need, you can decide on the type of fence that will best suit your short- and long-term goals. These days, issues of economy in material and labor have generally narrowed the choice to three general types: wire, electric, and rigid or semi-rigid.

These types can also be mixed to some extent, such as having an electrified high-tensile wire fence or a strand of barbed wire atop a wooden fence. For discussion purposes, however, it is handy to separate the types according to their main uses.

There are also any number of fencing solutions that use local materials, such as the rail fences that would have been familiar to Abraham Lincoln. These types are discussed at the end of this chapter.

WIRE FENCES

As industrialization began to make cheap, plentiful metal wire available after the middle of the nineteenth century, fencing was one of the many uses for which it was tried. It promised enormous advantages compared to the laborious process of making split-rail fences. In the newly opened western areas where wood and labor were scarce, wire fences appeared to offer a desperately needed solution.

No. 9 annealed iron wire was initially popular because it was commonly available, the relatively soft iron was easy to work with, and nine-gauge wire was considered "heavy" weight (see sidebar on how wire is measured). But the simple iron wire lacked strength, and cattle enjoyed rubbing against the smooth fence. The constant pressure loosened the posts and stretched or broke the wire.

It wasn't until the development of barbed wire that the use of wire for fence construction really became established, especially for cattle fences. Today barbed wire is still the first choice for effective, economical permanent fencing for cattle. The basic concept of barbed wire as a fence material hasn't changed much for over 100 years, but recent improvements include special coatings for resistance to corrosion and improved tools for handling wire. Barbs are also currently somewhat less aggressive than earlier designs because experience has proven that really dangerous barbs (such as used in military barbed wire) aren't needed

Form Follows Function

IF THE FENCE IS MAINLY A(N) . . .	THEN THE USUAL CHOICE IS . . .
Physical barrier	Wire fence (smooth high-tensile, barbed, or mesh)
Pain barrier	Electric fencing
Farm beautification element	Rigid fence
Environmental control element	Semi-rigid fence

The availability of cheap, plentiful local materials can make a big difference in the type of fence that's best for you. For example, an area with lots of industrial activity means a lot of surplus steel pipe and cable is available for a fence like this.

as much as was first thought and because ranchers are more concerned about the safety of their cattle's hides.

Barbed wire is also no longer the only effective wire for non-electrified fence. As noted above, barbed wire was originally developed to eliminate the problems with the stretching, sagging, and breaking caused by animals rubbing and pushing against the old non-barbed low-tensile wire originally tried as fencing. These days, a new generation of high-tensile smooth wire has come into use to eliminate the animal injury and wire handling problems of barbed-wire fence.

Early versions of the barbed-wire fence emphasized the pain that could be inflicted by extra-long barbs such as these. As wire fencing progressed, dangerous barbs were found to be less of an advantage than stronger wire between the barbs.

How Wire Gauge Relates to Thickness

When wire is described by a gauge number, rather than the actual diameter, the convention is the bigger the gauge number, the smaller the diameter of the wire. This may seem like a topsy-turvy way of numbering but there is a reason behind it. Wire is formed by pulling metal through a succession of smaller and smaller dies. Prior to the introduction of standard gauge sizes, the wire from the first drawing was called number one, wire from the second drawing was number two, and so on. The thinner the wire, the more times it was drawn through a die, so the higher the number.

This convention continued as gauge and die sizes were standardized in the nineteenth century. The current Brown & Sharpe American Wire Gauge (AWG) system was developed in 1856 by Rhode Island clock makers Joseph R. Brown and Lucian Sharpe. In this system, no. 2 wire (two-gauge) would be nearly as thick as a pencil, nos. 12 through 15 are typically used for barbed wire, and no. 44 wire is about the thickness of a human hair.

With the modern development of cheap, widely available tools to confirm actual wire diameter, the metric system and others have gone to the convention of describing wire by its actual diameter, such as 2.50mm barbed wire.

HIGH-TENSILE WIRE FENCE

In the 1970s, high-tensile wire drawn from high-carbon steel rod became widely and cheaply available. Compared to conventional low-carbon wire used in barbed wire or in baling or mechanic's wire, high-tensile wire maintains its toughness and flexibility when drawn to the same gauge. It can also be drawn to a thinner gauge and still maintain equal or greater strength than conventional fence wire. In addition, it is much harder to bend and strongly retains the shape to which it is bent. Because high-tensile wire is so springy, it tends to rapidly unreel by itself if the end of the wire gets loose while you're reeling it out. It also tends to loop into coils if the wire breaks once stapled to posts.

Fencing with straight strands of high-tensile wire was first developed in the 1970s in New Zealand, where two reasons drove its adoption.

First, stock dogs must be able to easily and quickly get through fences. Dogs are the main means of controlling and herding animals in New Zealand and Australia. While predator dogs can wriggle underneath barbed wire fences, getting through slows down stock dogs too much to accomplish their jobs. Second, the lighter weight per length means lower costs because steel is typically priced by weight. This is especially important in a country such as New Zealand, where industrial goods, such as manufactured steel, are generally imported and lead to high cost.

The many advantages of smooth, high-tensile wire fencing have now made it a much more popular choice for wire fencing.

High-tensile fencing has since spread throughout the world because of several other advantages:

1. It is safer for livestock, does minimal damage to animal hides, and is easily electrified to give better livestock restraint and predator protection.
2. It can withstand heavy pushing by livestock or low-temperature contraction without losing its elasticity. A typical 12.5-gauge wire has a yield strength of 1,600 pounds for each wire. Conventional low-tensile wire of the same gauge will yield strength of less than 500 pounds and break at less than 550 pounds force.
3. High-tensile wire is easy enough to handle during construction, provided you take a few simple precautions to prevent loose ends from getting free. It is flexible enough to bend, wrap, and tie in knots for splices, and it has no barbs to tear installers' skin or clothing.
4. It usually involves lower time and labor related to line posts because posts for high-tensile fencing can be spaced much farther apart. Corner posts do need to be more strongly braced to withstand the higher tension.
5. It is has a neat appearance when built and the high elastic limit of the wire keeps the fence looking neat for years thanks to reduced stretch and sag. The zinc coating on the wire turns white as it ages instead of the typical rusty brown of barbed wire.
6. Compared to an equivalent woven-wire mesh fence, electrified high-tensile fences for livestock can be constructed for about half the material and labor cost.
7. High-tensile wire fences can easily be electrified to construct temporary or permanent livestock enclosures. New fence-charger technology and fence construction methods now make long runs of electrified high-tensile fence easy to maintain.

WIRE MESH

For smaller animals that could easily slip under or between strands of high-tensile or barbed wire, woven-wire mesh fencing can be used. Mesh size ranges from the small, diamond-shaped opening of chicken wire to the large, square opening of page wire (also known as pig wire). A strand of barbed-wire or electrified wire can be added at the top to discourage animals from leaning over the fence or predators from climbing over. Woven mesh is available in either low- or high-tensile wire construction and with various kinds and grades of protection against corrosion: galvanizing, vinyl, or other finishes.

There are several different kinds of knot designs that hold wires together in the larger sizes of woven-wire mesh. Each kind is claimed to have various advantages in terms of keeping the fence tight, discourage animals from climbing the fence, and preventing damage to animals when they run into the fence. Ask your supplier about the type of knot used on the mesh

The merits of various types of wire fencing can be combined, such as in this common combination of mesh fencing topped with barbed wire to prevent animal pressure on the top of the mesh.

you're considering and get details on what advantages it may provide over other types of mesh.

The National Bureau of Standards within the U.S. Department of Commerce has standardized a set of eight standard heights for woven mesh fence wire. The specification for larger woven mesh is arranged as follows:

- The first number (eight or higher) represents the number of horizontal wires and the height in inches.
- The second number is the wire spacing in inches.
- Third number is the gauge of the intermediate wires.

For example, a wire mesh described as 939-6-12.5 means it has nine horizontal wires and a total height of 39 inches, with 6-inch wire spacing and 12.5-gauge main wires. The top- and bottom-edge wires are usually one gauge larger than the main wires. All woven-wire fencing comes in 330-foot rolls and can be high-tensile or standard wire.

Lighter-weight poultry or rabbit mesh is generally made with fine-gauge (e.g., 20-gauge) wire in a hexagonal (six-sided) mesh of 1- or 2-inch sizes. Standardized description of this wire is as follows:

- The first number is the height (width) of the wire.
- The second number indicates 1- or 2-inch mesh.
- The third number is the wire gauge.
- The last number is the length in feet of the roll.

For example, 48×1×20×25 poultry netting is 48 inches wide and has 1-inch hex size and 20-gauge wire in a 25-foot-long roll.

Sagging or bulging out is often a problem with mesh fences, especially with lighter mesh, such as poultry wire. To prevent it, a horizontal board can be attached at the top and/or bottom of the fence to provide a strong, secure attachment point for the mesh. The weight of wind-drifted snow loads can also severely stretch a woven-wire fence, so re-tensioning may be required every spring. If pressure from animals is causing the bulging, a strand of electrified wire can be installed to keep animals away from the mesh.

One of the main advantages of a woven-wire fence is that it is highly visible to predators, so it makes a good perimeter fence. A related disadvantage is that it tends to accumulate a lot of vegetation growing into it and wind-blown trash getting stuck up against it. Sheep may jam their heads through the mesh to get the grass on the other side, then be unable to figure out how to pull their heads back through. Unless the fence is regularly patrolled, the trapped animal may die of stress or become a convenient buffet for predators.

ELECTRIC FENCES

Energizing a metal wire fence with pulses of electricity has now become a safe and effective means of controlling the movements of farm livestock and unwanted intruder animals. A painful but non-lethal electric shock delivered on contact with the fence is usually enough to make animals avoid further contact. Since the shocks prevent continuous strong pressure on the fence, posts can be lighter and farther apart, which adds to the ease of construction for electric fences.

Electric fences can be made with bare, smooth high-tensile wire or various types of netting, tapes, or ropes that include some conductive wire. Electric fences using conductive tapes or ropes are especially useful for horses because they can see the line much better than they can a simple metal wire. Barbed wire is inefficient for electrified fencing because it loses a large amount of energy through the barbs.

Animals need some experience with receiving a shock from an electric fence before they will avoid it. With livestock, their normal habit of rubbing and pushing against any barrier quickly brings them enough experience to quickly learn avoidance. Other animals, such as deer and bears, need to be enticed into touching the fence so that they will receive the necessary training. This is done by putting attractive bait, such as scents or favorite foods, on the fence. When the animal touches the wire with its tongue or nose it receives a shock that delivers the required lesson in electricity.

Even with training, an electric fence may not deter a truly determined, hungry,

(Above) All electric fences should be prominently identified to prevent accidents. Signs are available wherever you buy electric fencing materials. *(Below)* An electric barbed-wire and mesh fence combination helps to contain several different types of livestock.

On its own, this simple pole fence might not stop many animals, but the electrified strand on top adds a lot of effectiveness.

or cornered animal. For this reason, electric fences may be beefed up by some level of physical barrier. One of the best combinations is electrified high-tensile smooth wire.

Some electric fences have all wires "hot," while some use alternating "hot" and "ground" wires. The choice is mainly determined by the electrical conductivity of the soil where the fence is installed.

In areas where there is reasonably good soil moisture, the soil will conduct electricity. All fence wires can be hot, and the jolt of electricity can be counted on to travel from the fence though an animal's feet and then through the soil to complete the circuit.

In more arid areas where there is low soil moisture, the soil conductivity is too low to complete a circuit. Ground wires are therefore installed right on the fence and are close enough that an animal is likely to encounter both a "hot" and "ground" wire and receive a shock.

The difference of grounding also has to do with what type of animals will be encountering the fence. Cattle have good contact with the soil due to their heavy weight, so an all-hot wire fence functions well. Other animals that are lighter or have smaller feet (e.g., foxes, raccoons, deer) have poor ground contact, so either a higher-than-normal charge is used to compensate for the higher resistance in the circuit or alternating hot/ground wire fences may be employed instead.

A stand-off electrified wire is hardly visible, but it can be very effective at keeping animals far enough away that they can't test the fence.

Electric fences should not be thought of as a "plug in and forget" fencing solution. Like any other type of fence, they require regular monitoring to make sure they are still functioning. If animals find a point of failure, they will not only immediately exploit it but also keep returning to that spot even after the fence is once again charged. The memory of being once able to break through at that point may overcome the shock they receive from the fence.

One of the key issues is that vegetation, either plants growing up or green branches falling down, may grow and make contact with the fence. Because green vegetation contains significant amounts of water, it conducts electricity well and may short out the fence. For this and other reasons, electric fencing should be patrolled regularly and fixed as required. Regular patrols will also discover the thousand and one other perils that afflict any fence, such as trees falling on it, wire breaking, posts loosening, and so on.

If a temporary electric fence is to be switched off for the season, the mesh, twine, or tapes should be also be rolled up and removed. Otherwise deer or other animals may break down and destroy the fence because it is not strong enough by itself to be a physical barrier.

RIGID FENCES

Fences with rigid horizontal members made of wood, plastic, or metal can provide an effective and attractive physical barrier. By adding a strand of electrified wire at some location where it's almost hidden from view

Rigid fences range from high-cost, high-maintenance types to simple types made with local materials.

(Above) Vinyl and other plastics are now being widely used in attractive, easy-to-erect fences with low maintenance requirements. *(Below)* The interior of plastic fence materials, as shown in the end of this section on hinge installation, shows how the smooth plastic boards are strengthened with hidden internal ribbing.

(e.g., inside the fence), the effectiveness of containment can be improved even more.

Fences made with wide, horizontal painted wooden boards tend to be the most expensive in terms of both construction and maintenance costs, and probably for that reason they are looked on as the most beautiful and prestigious. Farms encircled by long rows of gleaming white-board fence look prosperous and scrupulously maintained. Given today's cost of lumber and the labor involved with maintaining traditional painted wooden fences, they almost have to be!

Prefabricated steel panels can be quickly erected to provide a very strong barrier that's easy to move when the need arises.

To achieve the good looks without being burdened by the costs and maintenance chores, new materials are now being used in board-type fences. Specially coated metal or plastic boards or tubes are used to resist corrosion, cracking, fading, peeling, and other problems associated with wooden fences. The horizontal members use new methods of fastening to the posts to simplify construction and reduce maintenance. The posts themselves may also be designed for easier installation. In some designs, the horizontal members are round tubes, which provide far more resistance than flat boards to bending by animal pressure. The horizontal members may be bolted, not nailed, to the post to avoid problems with loosening, or they may slip into precast slots in the post. With bolt-on types, the hinged nature of the connectors allows the fence to easily follow contour changes. For anything more than a few feet of board-type fence, these new designs offer huge advantages while still preserving the desired classy looks.

Another popular and effective rigid fence is the type made by connecting prefabricated livestock panels made of

round or square steel tubing. This type of fence is very strong and easy to erect or take apart to move to another location. The fence can easily be enlarged as the budget for more panels becomes available. Although early versions tended to be quite utilitarian in appearance, newer models are being made with more attractive and durable finishes, such as powder coating in custom colors. Panel gates are also available to match this style of fence or to quickly add a strong gate in other types of fence.

(Left) Prefabricated, sag-resistant steel panel gates are easy to install in any type of fence. *(Below)* The loops at the bottom of these prefabricated steel panels help keep long gates from drooping.

RIGID FENCES WITH LOCAL MATERIALS

Although it's hard to beat the utility and cost-effectiveness of the dominant fence types described above, there may be occasions where older types of fences may be your choice. If you have an ample supply of trees and more labor than money available, a traditional all-wood fence can be effective for cattle and horses. Log or split-rail fences that rest on the ground can have low construction and annual maintenance costs if properly built, including provisions to prevent rotting at ground level. They can also add a considerable nostalgic value to your property. As with any kind of fence, a strand of electrified wire at some location where it's almost hidden from view (e.g., inside the fence) considerably adds to effectiveness while adding little cost.

If you can afford to wait, another option is to grow a fence by planting trees or bushes, preferably thorny ones close together, and then watering and aggressively pruning them to promote dense growth by encouraging an abundance of new shoots. These living fences can be quite attractive as a landscaping element, and they also serve as a windbreak and habitat for wildlife or small game, especially game birds. Until the fence becomes fully established, you

If wood is readily available at low cost, you might want to consider a traditional wood fence—it's attractive and can be effective for larger livestock, such as cattle and horses.

(Above) A farm encircled by a meticulously maintained white board fence whispers prosperity—as well it should, considering the cost of maintaining a traditional board fence. *(Below)* Wattle fences were developed thousands of years ago and can still work well. This current example weaves branches from a tough, thorny local shrub between inexpensive uprights of rebar.

can install a semi-permanent electric fence. Your state or province may also provide financial assistance for this type of fence via provision of free or low-cost plants for farm shelter belts.

To find out more about shelter belt programs in your area, contact your state or provincial agricultural extension and wildlife management departments. Alternatively, do an Internet search combining the words "farm shelter belt" and the name of your state or province, e.g., "farm shelter belt Wisconsin." The environmental control aspects of shelter belts are discussed in more detail in Chapter 9.

(Above) The classic split-rail snake fence quickly takes on a patina of age as the wood weathers. *(Right)* Recycled tires make a very durable solid fence. However, be aware that water pooling in the old tires makes an ideal breeding ground for mosquitoes. Before stacking, drill holes in the sidewalls for drainage.

CHAPTER

5

TOOLS AND EQUIPMENT

Because the clamshell shovel is stabbed into the soil, it's a little easier than the auger when working around small rocks.

Your fencing job will go a lot faster and more smoothly with certain tools made specifically for the tasks of installing posts and wire. A few basic shop tools will also be needed, such as a tape measure, hammer, saw, hatchet, drill, and bits.

BODY ARMOR

The first step is to equip yourself with personal protective and convenience gear, such as heavy-duty work gloves, boots, and clothing to protect your arms and legs from wood slivers and wire scratches. If you're working with barbed wire, especially if you are taking it down and rolling it up, expect to wear out several pair of gloves because those barbs do a great job of ripping and tearing. If working with a powered post pounder, use eye and ear protection. On all fencing jobs, bring plenty of drinking water, because this is hard work and getting dehydrated can lead to making bad decisions.

(Above) Even with today's wide variety of powered digging tools, the classic hand auger remains a useful and widely sold tool for fence construction. *(Below)* Push down and turn to dig, then carefully lift the auger to remove loosened soil.

A good tool belt is needed, as well, to keep hammers, pliers, and staples ready at hand. Stooping over to pick staples out of a tin can quickly get annoying. The sharp staples will quickly rip cloth belts, so get a heavy leather belt or pouch to hold the staples. Consider spending the money for a high-quality belt because if you're building any kind of fence, you'll eventually be doing maintenance where you'll appreciate a good tool belt. There will also be plenty of other jobs around the farm where you'll use it.

DIGGING TOOLS

For digging post holes, you'll need something that can go deeper than an ordinary spade can reach. For jobs where you're digging only one or two post holes, such as when replacing a broken post or adding a new post for a corner brace, hand-powered tools can be effective and easy to get to the site.

The basic hand-turned post-hole auger works well in soils that aren't too rocky or hard. It's also good for cleaning out or deepening the bottom of post holes dug with other tools. The hand auger by itself tends to drill a fairly small hole, so for placing a 6-inch-diameter post or larger, you will need to enlarge the hole enough to allow room for the tamping bar to fit in. A long, thin drain spade is good for this purpose and is also quite handy in the garden.

The pivoting blades of the clamshell shovel pinch the loosened soil and allow it to be lifted out of the hole.

The clamshell shovel is another widely used digging tool that allows you to reach

The pointed end of the digging bar (bottom) loosens soils and rocks. To make a suitable tamping bar, the digging bar at top has been modified with a wide, welded-on foot.

(Above) This long-bladed drain spade can reach far down to a post hole to enlarge it by scraping away the sides. *(Right)* A tractor-mounted post-hole auger makes short work of deep post holes. AGCO Corporation

quite a ways down a vertical hole. The handles are held together when digging, then moved apart to close the shovel and allow dirt to be lifted out. The Boston shovel is an older design that has a lever on the side to either move the spade into a vertical position for digging or horizontal position for pulling dirt out of the hole. The Boston shovel costs quite a lot these days and is rarely seen at hardware suppliers. If you see one at a yard sale or auction, it might be a worthwhile purchase.

For the stubborn rocks that you inevitably encounter partway down the hole, you'll also need a strong steel bar with a pointed end to work them loose. Another tamping bar with a blunt end is needed for tamping fill back in around the fence post once it is set into the hole.

Equipment rental shops usually have powered post-hole augers with a small engine atop the auger. If you do try one of these, be aware that when the auger hits a stone or hard soil, the handles tend to spin suddenly and vigorously, making it very hard for one person to hold. The weight of the digger also makes it quite tiring to move down a long stretch of fence line, as

Warning: Avoid this staged example of what not to do. When you're using the post-hole auger, keep everyone at a distance due to the potential hazard of entanglement.

often encountered in rural settings. There are better designs of powered auger available, but once powered equipment comes into play for fence post setting, most farmers seem to prefer post-hole augers that are mounted on the three-point hitch of a farm tractor.

POST POUNDING TOOLS

Thankfully, not all fence posts have to be set in pre-dug holes, because it is a lot of work. Even if you machine-dig the holes, the posts have to be tamped in. Cement could be okay for one or two posts, but it is far too expensive for a long line of posts.

Pounding posts straight into the ground is an effective and relatively easy option. The ground grips the posts well, and if you need to pound posts a long way down to reach dry, solid earth, you just use longer posts and keep pounding. A hand-powered post slammer or post maul can be used for pounding a few small posts. Another interesting option is an air-powered, hand-held post pounder.

For long lines of large wooden posts, the hydraulic post pounder is widely employed. The machines are either built on their own trailer and towed behind a tractor or mounted directly to the three-point hitch. Controls are provided to slide the pounding mast in and out and to level the mast so that that it is not necessary to drive into exactly the right position to pound a post. Smaller posts, such the metal T-posts used for electric fencing, can be pounded with the same kind of pounder as long as you take care not to apply so much force so fast that it bends or breaks the post.

Steel T-posts are an easy-to-install solution for rapidly erecting a fence.

The T-section of the post helps resist bending forces. The projecting bumps on the face of the post help hold wire or mesh in place during installation.

Safety consciousness is extremely important when using a powered post pounder. The heavy driving "hammer" of the post pounder can cause considerable damage to your fingers if they happen to get in the way, so be very careful about where you hold the post. Newer models of hydraulic post pounders have a post-holding device so you do not need to stand near the post when it's being pounded in. The post itself can also splinter under the driving force so work gloves, eye protection, and steel-toed boots are highly recommended when using the post pounder.

Older post pounders may have joints and bearings that have become quite sloppy through use and underlubrication. This makes the pounder harder to use, so be careful about letting yourself get so frustrated that you make bad decisions about safety. To keep your post pounder from getting that worn out, regularly lubricate joints and bearings with a good grade of high-pressure grease, preferably one with a high content of molybdenum disulfide (moly).

If using a towed post pounder, be sure to have the mast in its correct transport position before unhitching. If the mast is leaning back a little too far, its weight may make the trailer flip backwards when you unhitch it.

(Left) Let the slammer fall to drive the post into the soil and guide the post to a final plumb position. *(Right)* A hand-powered post slammer can be used to quickly install T-posts. Raise the slammer only enough that an inch or two of post remains within the tube.

(Above) The post maul (top) has a larger striking face and is made of relatively soft metal to avoid destroying posts while pounding. The sledge hammer (middle) is made for driving steel, not posts. Its smaller striking face is made of harder metal and will damage wooden posts. The splitting maul (bottom) is made for splitting wood and driving wedges into logs. *(Below)* The hydraulic post pounder sets long lines of posts in a short time.

Fatal Tractor Rollover While Setting a Fence Post

A sixty-year-old farmer was killed in a tractor rollover while using the front bucket of his tractor to push wooden fence posts into the ground. His son was holding each 6-inch post upright while the man, on the tractor, positioned the front bucket, half-filled with dirt, on top of the post. They were using the hydraulics of the tractor to push and/or ram the post into the ground, a procedure successfully completed with "thousands of fence posts" on their farm without apparent problems. In this case, however, the bucket slipped off the post and snapped downward with sufficient force and momentum that it caused the rear of the tractor to lift up. The wide-front tractor then rolled over on its right side down the slope, fatally crushing the farmer. There were no ROPS (rollover protective structures) on the tractor.

From Iowa FACE Report 03IA056

SPECIALIZED WIRE INSTALLATION TOOLS

Fencing pliers are a tool that's been developed over many decades to suit the needs of installing and maintaining barbed-wire fences. Since they are relatively cheap, buy several so you can keep pairs available in different vehicles and locations.

A regular hammer is a much better choice if you have a lot of staples to pound in because a hammer has a larger striking face, more weight for driving, and a more comfortable handle to hold. For occasional use, such as when it's the only tool with you while you are out riding the fences, the fencing pliers can do a decent job.

Fencing pliers are still the standard for maintaining barbed-wire fencing since they combine several functions into one easily carried tool.

This wire reel slips into the standard trailer-hitch receiver of a pickup truck.

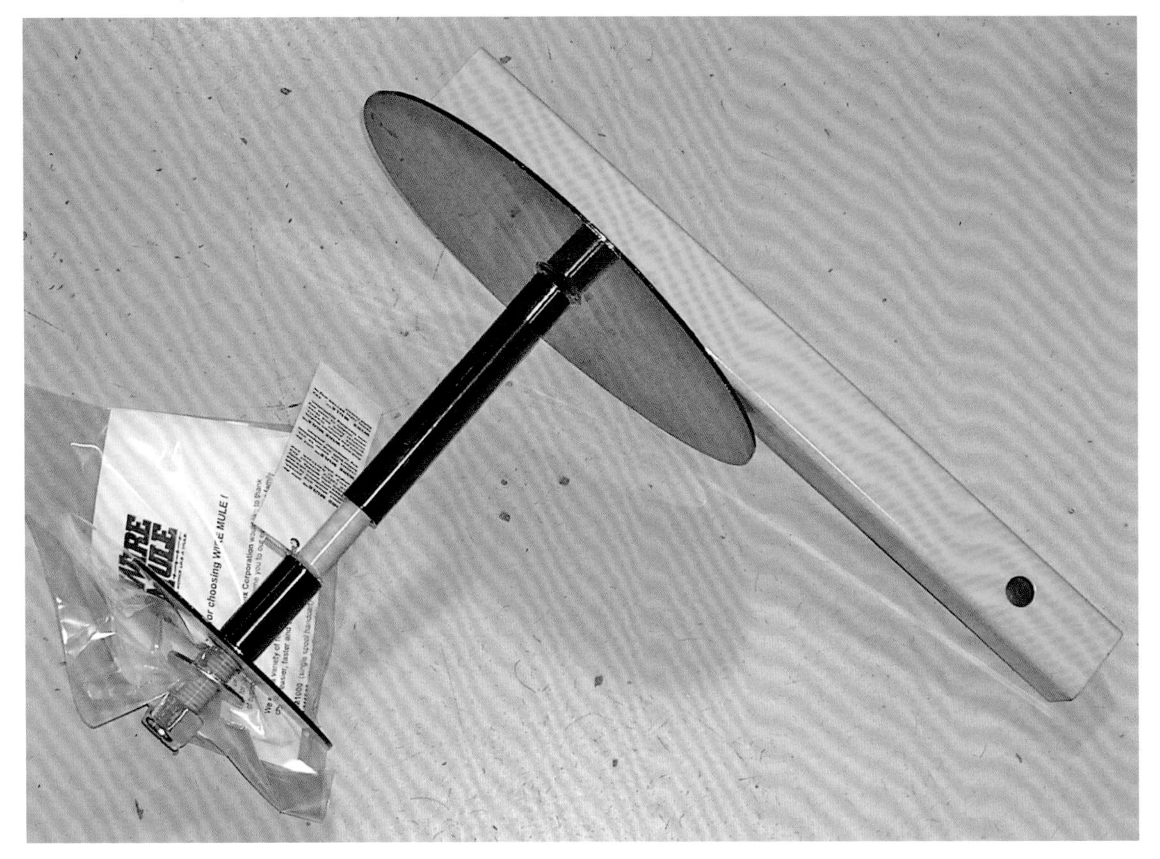

Compared to barbed wire, high-tensile wire is much harder to cut, so the high-tensile wire cutter addresses that need for more cutting power. You'll wear yourself out trying to cut high tensile wire with ordinary wire cutters and quickly ruin your tools. To keep the pliers sharp for a longer period of time, try to keep the edges at right angles to the wire when cutting. Keeping any cutting tool blades free from rust is also important for maintaining the ease of cutting, so wipe down the pliers after use and before storage and give them a light spray with WD-40 or similar lubricant. Mechanical linkage on the high-tensile wire-cutter jaws allows exertion of up to twenty times the force applied at the handles, and you'll need it because high tensile-wire is very strong.

The Grippler tool pulls wires together into a splice. When tool tension is released, the connector grips the wires tightly without losing any wire strength.

Hand reels simplify the handling of electrifiable twine, rope, or tape. Some types have a neck strap or a hook to attach to the racks on an ATV.

ELECTRIC FENCE TROUBLESHOOTING TOOLS

As with any fence, electric fences need periodic monitoring to make sure they remain capable of their intended job. Along with checking the obvious physical condition of posts and wires, checking the electrical functioning is also necessary. It's not enough to know that the fence energizer is switched on and connected. You also need to know that all the "hot" wires are along their full length and just how hot they are. Areas of the fence that deliver less than about 4,000 volts will not be very effective against excluding predator animals, such as coyotes or dogs. Sheep may start testing the fence if voltage drops below 3,500 volts. A charge below 3,000 volts starts becoming ineffective for keeping cattle and horses away from the fence.

To monitor the electric fence's condition, you'll need two relatively inexpensive tools: a voltage reader and a small, battery-powered transistor radio. The voltmeter can confirm that your energizer is charging the fence, that the grounding is adequate, and that power is reaching the end of the fence. A shop-type multimeter does not work well for electric fences. Instead, specialized voltage testers for electric fences are available at the same place you buy other electric fencing products. The three types, in order of expense, are as follows:

1. The basic positive-voltage reader (about $5) lights up when it detects a hot wire. It does not tell you how hot the wire is, only that some charge is in the wire.
2. The more advanced multi-LED voltage reader (about $15) has a series of five lights that illuminate at particular voltage levels (typically 1,000, 2,000, 3,000, 4,000, or 5,000 volts) on the line. In many cases this go/no-go information is sufficient to check if the fence is functioning correctly.
3. The most advanced maintenance tool is the digital voltage reader (about $40 and up). It displays the actual charge, accurate to the nearest 100 volts, up to a maximum charge of 9,900 volts. The finer output reading enables you to narrow down the problem area. Better models also indicate the direction of the fault from the point at which you are testing–something that is very good to have.

A transistor radio can help locate cracking or arcing insulators that are otherwise hard to detect. That's because the electrical arcing in a faulty insulator gives off radio-frequency energy, a smaller-scale version of the burst of radio static given off by lightning. If you tune the radio between stations and then approach a faulty insulator, you'll hear clicking from the arcing going on in the faulty insulator.

Various types of detachable grips are available for tensioning fence wires. In general, some type of wedging action grips the wire without causing nicks or kinks. In this example, a sliding collar grips the wire against the conical wedge in the center.

Monitoring charge in electric fences starts with inspecting the energizer and the grounding system. Next, check the wires connecting the energizer to the hot wires on the fence. Follow the hot wires to the end of the fence. If you find faults, refer to the troubleshooting chart on page 81 for probable causes and effective solutions.

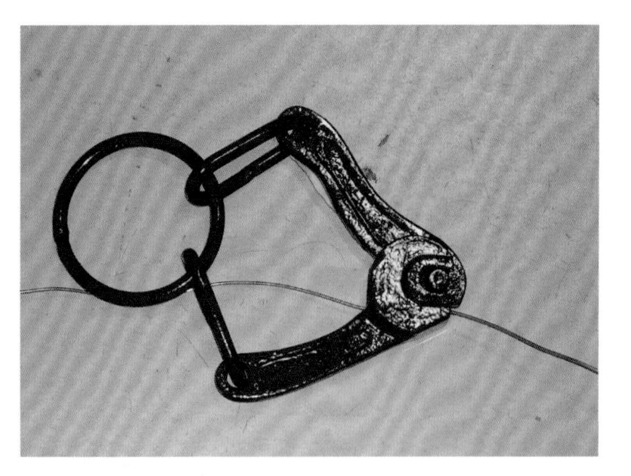

(Left) Another common type of wire gripper shown in the open position allows the wire to slide freely between the jaws. *(Right)* When the gripper is closed by applying a pulling force to the ring at left, the jaws move together to grip the wire.

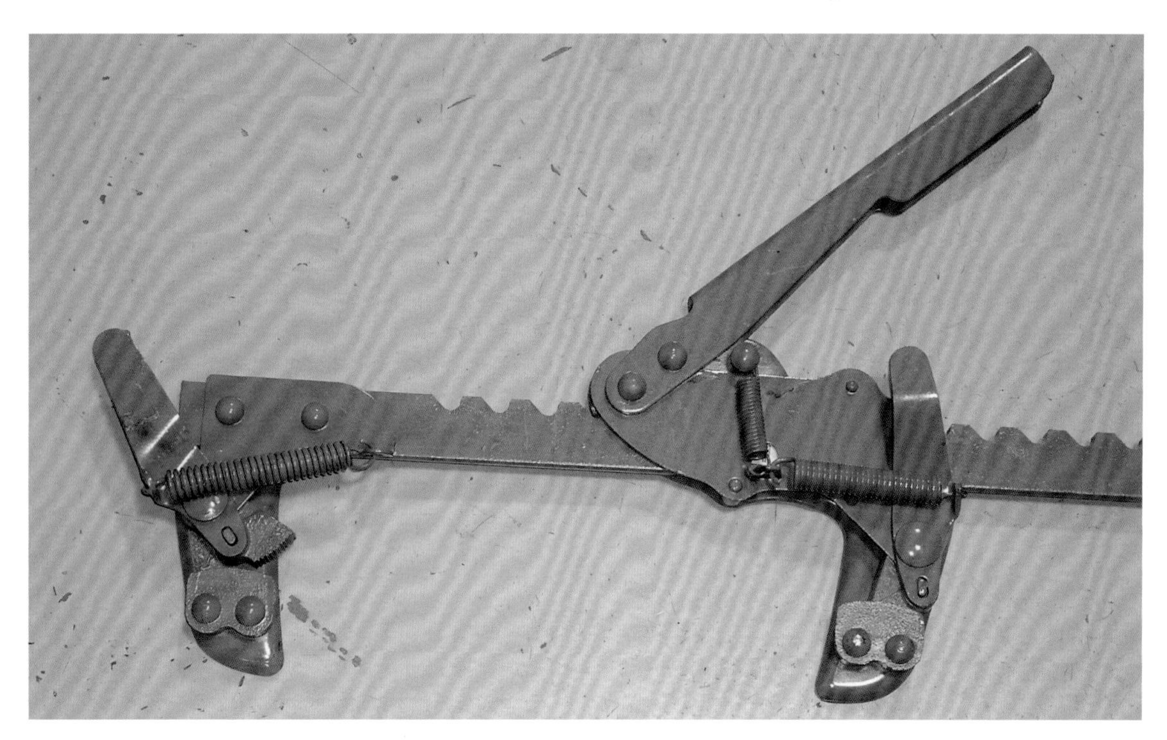

A typical wire-tightening device can grip both ends of a splice or repair. Moving the handle at the top pulls the jaws together. The more pulling force that's applied, the more tightly the wire is gripped by the spring-loaded jaws at each end of the tool.

Electric Fence Troubleshooting Guide

Condition: Energizer disconnected from fence, voltmeter reading zero at energizer output terminals
Cause and remedies
- Energizer switched off: switch on
- Main power off: check plug with another appliance
- Blown fuse on input circuit: replace fuse
- Batteries dead/discharged: replace or recharge
- Corroded battery connectors: clean and retest
- Output terminals corroded: clean and retest

Condition: Energizer disconnected from fence, voltmeter reading zero at energizer output terminals
Cause and remedies
- Energizer switched to low setting: switch to high setting
- Weak batteries: replace or recharge
- Corroded battery connectors: clean and retest
- Output terminals corroded: clean and retest

Condition: Energizer output good but zero voltmeter reading on fence
Cause and remedies
- Ground-return wire faulty: check wire resistance, clean, and tighten connections
- Energizer-to-fence wires faulty: check wire resistance, clean, and tighten connections
- Broken live or ground-return wire on fence: check fence voltage readings

Condition: Energizer output good but voltmeter readings low at many places on fence
Cause and remedies
- Energizer switched to low setting: switch to high setting
- Green weeds, grass contacting fence: remove vegetation
- Energizer inadequate for length of fence: replace with higher-output unit
- Ground system inadequate: follow energizer manufacturer's guidelines for ground system construction
- Soil dried out: fence may require higher-output energizer to compensate

Condition: Energizer output good but voltmeter readings low at one place on fence
Cause and remedies
- Broken wire: repair
- Broken or disconnected jumper wire: repair
- Short-circuit across wires: find and remove
- Disconnected or deteriorated ground rod: repair
- Green weeds, grass contacting fence: remove vegetation

Condition: Energizer output good but voltmeter reading on one wire higher than another or no reading from one live wire to ground-return wire or soil
Cause and remedies
- Broken or disconnected fence wire: repair
- Broken or disconnected jumper wire: repair
- Broken or disconnected ground wire: repair
- Broken or faulty insulator: check with transistor radio as described on page 79, replace if faulty
- Ground rod deteriorated: repair

Condition: Operating the fence causes radio, television, or telephone interference
Cause and remedies
- Ground system inadequate: follow energizer manufacturer's guidelines for ground system construction
- Antenna too close to fence: relocate antenna
- Fence parallel with antenna wires or telephone wires: fence may have to be changed to non-electric type if interference with electronic devices is unacceptable.

CHAPTER

6

CHOOSING AND INSTALLING FENCE POSTS

Wooden posts can be "dressed up" by adding a cage of wire mesh, then filling the cage with local stones.

Wood is the most commonly used material for agricultural fences, and the type of wood post you choose will make a major difference in ease of construction and longevity of the fence. Steel T-posts are lighter to work with and easier to drive into the ground, but they are more expensive and not as strong as wooden posts. The steel used in T-posts is quite soft, although the T-section does improve rigidity. Steel posts are often useful for quickly erecting a temporary fence that is not subject to large loads.

Concrete posts are very strong, durable, and environmentally safe, they can be produced on the farm, and, if properly steel-reinforced, they can be driven with a standard post driver. The only modification necessary to the pounder is a rubber cap used to protect the top of the post from chipping. However, concrete posts are very heavy to handle and require casting-in of some sort of insert to allow attachment of wire or boards.

To reduce costs and labor in forested areas, you may be tempted to use living trees as posts. It's not the best solution because as trees sway in the wind, it tends to loosen any wires or boards attached to them. The tree will also grow over any wire nailed directly to it, which makes subsequent re-tensioning of wire more difficult. If you do use living trees as posts, nail a length of lumber (e.g., rough cut 2×4) to the tree, then attach the fence wire or board to the "nailer." Attach the nailer to the tree using Ardox (spiral shank) spikes to maximize holding power. Do not drive the spikes fully into the tree and board. Leaving an inch or two of the nail exposed allows room for the nailer to shift outward as the tree grows in diameter. If you wrap the wire completely around the tree, use nailers all around the tree. This keeps the wire from choking and killing the tree as it grows.

WOOD POSTS

The simplest and, often, best choice for wooden posts is the type of round pressure-treated post available at farm supply outlets. The pressure-applied preservative on these posts is very effective at resisting rot and fungal attack. Post tops are chamfered to resist splintering when used with a post pounder. While the ends are pre-pointed to make driving easier, the quality of the pointing is often somewhat poor, with not all sides being equally sloped. If you can, select posts that have points as evenly made as possible. This will make post pounding a lot easier.

Local wood can also make good post material. While they involve more work and time to prepare, posts made from farm materials may also involve less cost and be more suitable to local conditions. Split cedar posts provided very satisfactory

Ideally a post made for pounding would be equally sharpened on all sides to keep the post straight while driving it into the ground. This example shows what you are more likely to deal with.

Typical Life Expectancy (Years) of Untreated Wooden Posts

From "Selecting Preservative Treated Wood," University of Minnesota Extension Service

VERY DURABLE	
Eastern red cedar	30-plus
Redwood*	10 to 30
Western red cedar	10 to 25
DURABLE	
White and burr oak	10 to 15
Northern white cedar	5 to 15
MODERATELY DURABLE	
Tamarack	8 to 10
Red oak	6 to 8
Douglas fir	4 to 6
NOT DURABLE	
Red and jack pine	2 to 6
Aspen (poplar) and cottonwood	3 to 4
Ponderosa pine	3 to 4
White birch	3 to 4
Spruce and balsam fir	3 to 4
Basswood	Less than 5
Maple	2 to 4
Ash	Less than 5
Willow	Less than 5

*Although tests at the Forest Products Laboratory in Madison, Wisconsin, show that redwood durability can be good, it is at best quite variable. Their recommendation is treatment of redwood whenever it is used in ground contact.

service life in the past and can still do so where available at reasonable cost. Osage orange heartwood is the most decay-resistant of all North American timbers and is immune to termites. These posts have been described as being able to "outlast the hole." On the other hand, woods like poplar or cottonwood are poor choices for posts unless that's all you have, you need a no-cost fence fast, and you can accept replacing the post with better materials within a few years.

SEASONING LOCALLY MADE WOODEN POSTS

If you're using wood cut on your farm as a post material, there are some advantages to seasoning (drying to remove excess water) before setting posts in the ground. Seasoned posts do not check or crack as much, are lighter to handle, and generally hold staples in place longer. Most methods of farm-treating posts require seasoning prior to preservative application. Unseasoned posts may check or crack after treatment, exposing untreated wood to decay caused by fungi and insects.

To season your own posts, remove bark and stubs of branches. Pile posts loosely on supports at least 1 foot above the ground in a well-ventilated location. Allow 60 to 90 days in summer and 120 to 180 days in winter for seasoning. Checks and splits can be minimized by shedding or covering the stacked posts so that they do not dry too rapidly.

Since much of today's commercial lumber comes from tree plantations, you are likely to find many posts or timbers made from small trees with a lot of sapwood. This makes post treatment crucial for decent service life.

WOOD POST DECAY PREVENTION

Once wood posts are set in the ground, they are subject to feeding by fungi, insects, and other soil organisms. The result can be weakening of the post to the point where it can no longer support fencing materials. The more moist and oxygen-rich the soil, the more rapid the decay. Optimum conditions for fence post decay usually occur about a foot below the soil surface, down where you don't generally notice it until the post breaks.

Some woods, such as cedar and Osage orange, naturally resist decay in the soil. For more common and less expensive woods, application of chemical treatments reduces decay and increases post service life. For example, in Missouri conditions, untreated posts made from common woods, such as southern pine, hickory, red oak, sycamore, poplar, or cottonwood, typically only provide two to seven years of service. Once those

This post was loose, and upon being pulled it is easy to see why: soil organisms in the upper layers of soil have consumed much of the wood fiber. In time, this post would have been weakened so much it would easily snap off at ground level.

same woods are pressure-treated, expected service life jumps to a 25- to 30-year range.

The application of wood preservative by pressure treatment is by far the most effective method of decay prevention, but soaking and brushing methods can also be suitable for preserving posts made on the farm. Farm and building supply stores have various products available. Find out what works best in your area and get recommendations on how to apply it effectively and safely.

In the past, fire was used to char the surface of the post where it was in contact with the soil. While it is inexpensive and

How Rot Treatment Extends Life Expectancy of Wood Fence Posts

From "Fences for the Farm," The University of Georgia College of Agricultural and Environmental Sciences Cooperative Extension Service

TYPE OF WOOD	TYPICAL YEARS OF SERVICE	
	Untreated	Treated
Ash	3 to 7	10 to 15
Aspen	2 to 3	15 to 20
Bald Cyprus	7 to 15	20 to 25
Balsam Fir	4 to 6	10 to 15
Basswood	2 to 3	15 to 20
Beech	3-7	15
Birch	2-4	10 to 20
Black Locust	20 to 25	Not necessary
Box Elder	2 to 7	15 to 20
Butternut	2 to 7	15 to 20
Catalpa	8 to 14	20 to 25
Cedar	15 to 20	20 to 25
Cottonwood	2 to 6	10 to 15
Douglas Fir	3 to 7	15 to 18
Elm	4	15
Hackberry	3 to 7	10 to 17
Hemlock	3 to 6	10 to 25
Hickory	5 to 7	15 to 20
Honey Locust	3 to 7	10 to 20

TYPE OF WOOD	TYPICAL YEARS OF SERVICE	
	Untreated	Treated
Larch	3 to 7	10 to 20
Maple	2 to 4	15 to 20
Oak (red)	5	15
Oak (white)	10	15 to 20
Osage Orange	20 to 25	Not necessary
Pine	3 to 7	25 to 30
Red Cedar	15 to 20	20 to 25
Red Mulberry	7 to 15	15 to 30
Redwood	10 to 15	20 to 30
Sassafras	10 to 15	20 to 25
Spruce	3 to 7	10 to 20
Sweetbay	2 to 6	10 to 20
Sweetgum	3 to 6	20 to 30
Sycamore	2 to 7	20 to 25
Tamarack	7 to 10	15 to 20
Tupelo (black)	3 to 7	15 to 20
Willow	2 to 6	15 to 20
Yellow Poplar	3 to 7	20 to 25

(Left) This slice of soil from the side of a post hole shows the rapid change from black, biologically active topsoil to lighter-colored clay that packs better. *(Below)* The wide foot of the tamping bar (right) is great at packing the soil, but if the post gets too close to the side of the hole, you may need to use the digging bar or turn the bar over and use the narrow end for tamping.

relatively effective, it is also very slow, so it would only be practical for setting a few posts unless you have time, as well posts, to burn. Products like coal tar, crude oil, furnace oil, or diesel fuel may have an impressive stink but actually have little value as wood preservatives for posts.

ORGANIC ALTERNATIVES TO CONVENTIONAL POST TREATMENTS

Treated lumber, including fence post material, is not allowed under the Final Rule of the National Organic Program (NOP). For posts and fence boards in contact with soil, crops, or livestock, current and prospective organic producers have some options for wood treatment if options like metal, concrete, recycled plastic, plastic/wood composite, or untreated wood, are not suitable.

Osage orange, redwood, Eastern red cedar (juniper), Western red cedar, black locust, and bald cypress are woods more naturally inclined to resist decay than others. However, there can be a wide range of life expectancy even for woods generally known to be highly durable. Posts made from heartwood will make more durable posts than sapwood, which is the lighter-colored wood between the bark and the darker inside core of the log. As new sapwood forms in outer rings, sapwood closest to the center of the tree dies and becomes heartwood.

Approved alternative wood preservatives can help extend the life expectancy for most wood. Borates (boric acids and borax) are not considered suitable for unprotected outdoor use, such as for fence posts or poles, but they are suitable for most building construction purposes.

Ammoniacalcopper citrate (CC) is recommended by sellers for use in treating fence posts and grape stakes. Copper azole (CBA) is a wood preservative formulation used in commercial preservatives, such as Wolman® E preservative. Alkaline copper quaternary (ACQ) ammonium is a rot, decay, and termite attack preventive used in commercially treated wood.

A wood-preservative recipe was developed by the USDA's Forest Products Laboratory (FPL) to protect wood used above ground for up to 20 years. The preservative was not tested for effectiveness when in contact with soil (e.g., on posts), but it is said to be nontoxic when prepared with ingredients listed on the National List of Allowed and Prohibited Substances as NOSB-approved or as approved by your organic certifiers. The recipe may not prove sufficient where posts face frequent exposure to moist soils, such as the Pacific Northwest or the southeastern states, where an addition of a copper-based product may be needed.

Wood Preservative Developed by the Forest Products Laboratory

Recipe from "Organic Alternatives to Treated Lumber," Appropriate Technology Transfer for Rural Areas (ATTRA), July 2002

INGREDIENTS:

- 1½ cups boiled linseed oil
- 1 ounce of carnauba or wood rosin wax, provided they contain no prohibited substances
- Enough solvent (distilled pine tar, mineral spirits, paint thinner, turpentine, citrus thinner, or whatever is approved) at room temperature to make the total volume of the mix 1 full gallon

DIRECTIONS:

- Melt the wax over water in a double boiler (do not heat over a direct flame). Away from the heat source, slowly stir the melted wax into the solvent while vigorously stirring the solvent.
- Add the linseed oil and continue to stir thoroughly until fully blended.
- Apply the treatment by dipping the untreated lumber in the mixture for three minutes or by brushing a heavy application across the wood's grain and on the cut ends of the lumber.
- Wood can be painted when the treatment has thoroughly dried.
- The mixture may separate and settle out when cool. To prepare cooled mixture for use, warm it to room temperature and stir vigorously.

CAUTIONS:

- Flammable solution: do all mixing outdoors and keep firefighting equipment handy.
- Wear gloves and eye/face protection while mixing and applying.
- Avoid breathing the vapors.

Copies available by calling 800–346–9140 or visiting www.attra.ncat.org.

INSTALLING POSTS

Before installing any horizontal members, such as wire or boards, the fence posts need to be set firmly in place. The sequence begins with the start/end posts at corners, continues through posts set at any changes in direction, and finishes with line posts set on the straight lines in between.

Two of the first decisions to be made are what size of post should be used and how far apart they should be placed. The answers depend a great deal on what the fence is to enclose or exclude and on the condition of local soils. The best solution to suitable post size and spacing is usually obtained by observing and measuring existing fences in your area. Based on that preliminary survey, you can decide to increase or decrease the post size, spacing, and setting depth according to whether you need a stronger fence.You can also reduce sizes, spacing, and depths to achieve a more economical design.

Posts for high-tensile fence do not need to be as close together as they are for a barbed-wire fence. In fact, the effectiveness of high tensile-fencing can be increased (and the cost lowered) by spacing the posts farther apart. They can be placed up to 100 feet apart on rangeland and 50 feet on pasture.

In general, 4- to 6-inch-diameter wooden posts will be suitable for line posts on most wire fences with posts 6 inches or larger for corners and braces. About one-third of the length of the post is set into the ground; a typical 6-foot wooden post would have 2 feet in the ground and 4 feet available for attaching the fence. Be careful about using smaller posts because strength drops off dramatically as diameter decreases.

Availability of local materials also plays a part in design. If you can get large-diameter, heavily rot-proofed posts at a bargain price, such as due to sell-off of old power poles or railway ties, installing them at corners will improve strength with what may be only a slight increase in labor requirements for digging the holes. The extra strength will pay off for decades, while the extra labor will be forgotten after a short time. Posts can be either tamped into a pre-dug hole or forced into the ground through pounding, pushing, or vibration by hydraulic machinery.

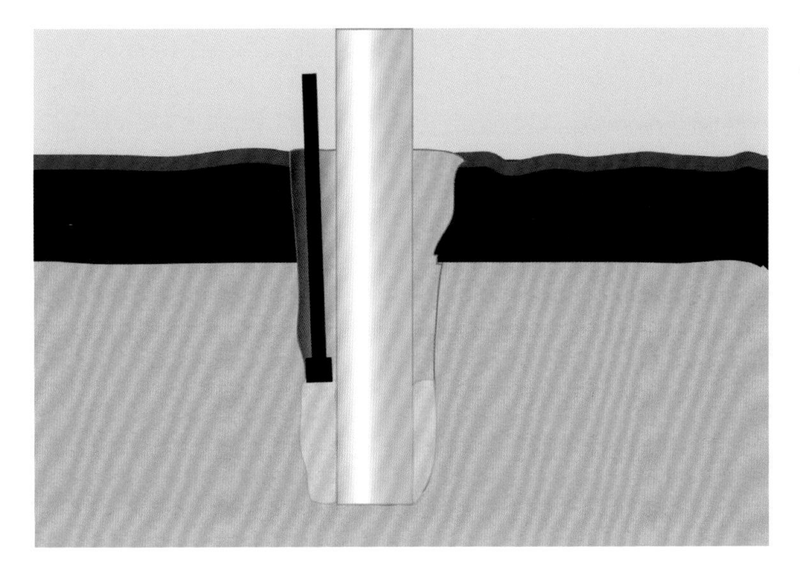

Posts stay steady longer when you make an extra effort to get the bottom third tamped in really tight.

An easy way to keep posts plumb while tamping is to line them up with trees some distance away. Eyeballing posts to be parallel with tree trunks can save the need to pick up and put down a level.

Determining Post Breaking Point

Post diameter	Post breaking force, steady load applied at 4 feet above ground
2½ inches	225 to 250 pounds
3 inches	385 to 430 pounds
4 inches	915 to 1,025 pounds
5 inches	1,785 to 2,000 pounds
6 inches	3,080 to 3,455 pounds

From "Fence Posts: Material, Installation and Removal," British Columbia Ministry of Agriculture, Food, and Fisheries

DIGGING AND TAMPING

This is often the easiest choice if you're installing one or two posts, particularly large posts, or if you don't have access to a tractor and post pounder. Post holes can be dug with a shovel and bar or augered out with power machinery.

Compared to posts that are pounded in, posts that are dug and tamped in are less resistant to pulling out. Their sturdiness can be increased by installation with the larger end in the hole, so there is a wedging action against the sides of the hole. Digging and tamping in more than one or two posts is a lot of hard physical work and must be done well to resist uplift from tension of the wires above or heaving of the soil below.

Setting a stone at the bottom of the post hole is not recommended because the post is more likely to be heaved upwards by the action of water and frost. Stones can be successfully used as fill on the sides of the post to help resist loads that pull the posts against the sides of the hole.

When tamping around a post, the best results are achieved by filling in the hole a few inches at a time and then firmly ramming the fill material into place. The more firmly the bottom third of the post is tamped in, the stronger the final setting will be. Use a steel or wooden bar with a larger-diameter "foot" that packs the material.

If you are tamping the post with soil that came out of the hole, use the driest, most clay-rich material at the base. Clay soil tamps down well, resists moisture infiltration, and can withstand plenty of force, which is why basements and other engineering works are only constructed once the site has been scraped down to the clay layer.

An even better method of tamping posts solidly into holes involves using road crush gravel. Road crush is the relatively clean but unsorted mixture of sand, pebbles, and stones spread on gravel roads. Because it's been less processed, it's quite inexpensive compared to more finely sorted material, such as sand or pure gravel. Filling and tamping with road crush also has the advantage of keeping the base of the post relatively dry. Water can easily drain through the gravel and away from the base of the post. This extends the life of the post by reducing wood rot. Utility companies use this material for securely setting large power poles.

It is possible to use concrete to set farm fence posts, but there are many reasons why it is not a preferred method. First of all, concrete is relatively expensive and rather messy, as well as being caustic to your skin. Second, being set in concrete can actually shorten the life of the post because even if the top of the concrete plug is properly sloped, water seeps in between as the post dries and shrinks. Third, it makes the fence post very hard to remove and dispose of when you have to move the fence. You may think that you want the post to be firmly planted forever, but always think ahead to save time and effort. Experience indicates that posts will need to be removed as the farm operation changes. Having a massive concrete plug stuck on the bottom of a post makes removal difficult if you don't have equally massive machinery.

POUNDING

Pounding posts into place makes for much quicker work when there is a long line of posts to be done and there is sufficient room beside the fence line to drive the tractor and post pounder. Pounding is a preferred method for low-lying ground where post holes fill up with water before you can finish digging them. Just be sure to use posts that are long enough to reach down to solid ground.

If you need to install a post that is so long it exceeds the post pounder's capacity, you can dig a hole just deep enough to achieve clearance, then finish driving the post once it is set in the hole. The driven post will be held firmly enough that the top bit of fill will not need to be tamped as tight as a hand-pounded post.

If you have a tractor but don't own a post pounder, units are often available for rental. The per-day charge is usually a considerable saving over renting the unit for only a few hours, so if you rent a post pounder, plan your work to make maximum use for the money. Splitting the rental with a neighbor who also needs to do a bit of fencing may also be a way to defray the cost.

The tractor-towed post pounder is basically a large weight that is lifted slowly with hydraulic power, then dropped rapidly to drive the post in like a hammer driving a nail. As with a hammer and nail, start with a few firm taps to get the post going straight, and only then proceed with full pounding power. If you are pounding in very large-diameter posts or if the ground is particularly hard, using an auger to create a small pilot hole can make pounding easier.

Safety consciousness is extremely important when using a powered post pounder. When you hold the post make sure your hands won't get in the way of the hammer of the post pounder. Newer models of hydraulic post pounders have a post-holding device so that you do not need to stand near the post when it's being pounded in. Work gloves, eye protection, and steel-toed boots are highly recommended when using the post pounder.

Regularly lubricate your post pounder's joints and bearings with a good grade of high-pressure grease, preferably one with a high content of molybdenum disulfide (moly), to keep it in good and safe working order.

When pounding posts, lightly stretching the lowest strand between the start/end posts provides a clear indication of where line posts are to be pounded.

With the post point set in place, the pounder mast is adjusted to start the post off as close to plumb as possible.

Post-pounder controls, especially on older machines, are often not clearly marked, but they generally follow a similar pattern. The hydraulic control handle by itself at the extreme right raises and lowers the weight for pounding. The three control handles grouped at left are for adjusting the pounder position in vs. out, leaning forward vs. back, and leaning in vs. out.

The pounder is pulled in and straightened up for transport between posts.

The pounder is slid out and plumbed in preparation to start pounding a post.

Due to unevenness in the post point and variations in soil, the post often goes crooked after one or two hits. Keeping it straight with hearty shoves and occasional realignment of the pounder mast is usually needed. Never put your fingers on top of the post!

You may run across soil conditions where normal pounding or dig-and-tamp methods just won't work and above-ground posting methods are needed.

With a towed post pounder make sure the mast is in its correct transport position before you unhitch it from the tractor. If the mast is leaning back a little too far, its weight may make the trailer flip backwards when unhitched.

ABOVE-GROUND POSTING

If hard, rocky, or steeply sloping soil makes pounding or dig-and-tamp methods impractical, above-ground structures can stand in for posts. The key point when using above-ground posts is to make them heavy enough to withstand forces exerted by wires or by animals pushing and rubbing against the posts. Wooden structures weighed down with stones are a typical solution because they have sufficient weight for stability, plus some wood to which wires can be stapled. By using local wood and rocks, cost can also be kept to a minimum.

A heavy buttress can be placed above ground to make a firm attachment point for fence wires.

SETTING THE START/END POSTS

Posts at the beginning or end of the fence need extra strength because they are subject to imbalanced forces. These posts are pulled hard by wires from only one side, unlike line posts, which have even tension from both sides. Some of the force imbalance will be canceled out by installing diagonal braces, but that does not cancel out the need to set start/end posts as firmly as possible in the ground. Start/end posts should be of a larger diameter than other posts and set more deeply into the ground.

In wetter areas, a "turner" is attached to the bottom of the post. This small crosspiece notched and nailed into the base of the post resists rotational forces caused by tensioning the wire. It also makes it harder to pull the post out, so observe local practices to see if turners are necessary for your conditions.

POSTS AT BENDS

If you take a look down the line from the start post to the end post and see that the proposed fence line bends more than one or two widths of a post, the posts to

This closer view of the buttress shows how the tripod is weighted with stones to prevent movement. If needed, additional wrappings of wire on the tripod joints could improve structural durability.

Any significant bend in the fence line adds additional strain on posts at the bend.

be located at bends need special kinds of installation. The degree of deflection can be determined by taking sights with a compass or by measuring the distance off line, as shown in the drawing at right.

For a deflection up to 20 degrees off a straight line, you can avoid setting braces by using a larger-diameter post leaned 4 inches toward the side where the wires will be attached. The post should also be set into the ground as deeply as a start/end post–at least 3 feet.

For a deflection between 20 and 60 degrees off a straight line, you can avoid setting braces by setting posts 4 feet apart and making a series of 20-degree deflections using the technique described above. For a deflection more than 60 degrees off a straight line, you'll need to install a start/end post complete with braces.

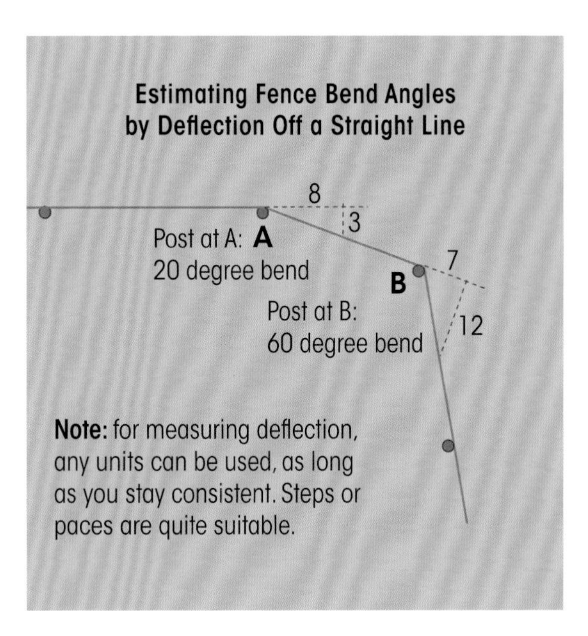

For judging the degree of bend, you can use a compass or make a judgment based on the amount of deflection off a straight line.

An extra-heavy post at a bend is always a good idea because post strength increases exponentially with increasing diameter.

LINE POSTS

With start/end and corner posts set, line posts can be set in between. Since line posts are subject to even tension loads from both sides, they can usually be of a smaller size and not set as deeply in the ground as the corner posts.

If the straight fence line stretches for a long distance, the durability of the fence will be improved by having braces at the line posts where the wire is to be tensioned. Ideally, barbed wire should be tensioned every 660 feet (1/8 mile), while high-tension smooth wire fences can be tensioned every 1,320 feet (1/4 mile).

On reasonably level ground, establishing a straight line to place the line posts is simply a matter of stringing out the bottom wire and pulling it tight at one end. If that's not practical in your situation, you'll need to establish points on line so you can visually line up individual posts. Ideally this is a two person job, with one person sighting from one corner post to the other and giving alignment directions to the second person. Light lath can then be pounded in on-line and the posts can be sighted in once a line of lath is placed. Experience indicates that as the distance between the persons stretches beyond vocal range, it can become

Inline braces that are closely spaced (at center) improve the effectiveness and durability of wire fences by providing more places to tension the wires.

To run a line between the foreground post and the end post (red arrow), pick a distinctive backsight that's on line and behind the farthest post. In this case, it's a black tree with a dead tree just to the left.

A stake was placed approximately on line by walking toward the far post and trying to keep it lined up with the backsight. As expected, the center stake wasn't quite correctly placed. But the correct line does fall right between the left and center stake. If it were even farther off, knowing the outside stakes were 1 and 3 feet off, respectively, would help judge how far to correct the line.

A new line stake is placed between the left and center stakes. A quick check at the starting post shows it is now right on line.

With a midpoint stake now established, you can line it up with a start/end post to establish any remaining points on line as needed.

somewhat problematic to give directions, which leads to irritation, frustration, and harsh words being exchanged about the quality of the job being performed. This is especially noticeable when a husband and wife team is doing the job. Two-way radios or cell phones with a walkie-talkie function are a great help and enable communication.

The work tends to go much more smoothly and efficiently if one person can do the job of getting the posts in a straight line. A good-quality GPS receiver can do the job quite well nowadays and has plenty of other uses around the farm and for recreation, so that's something worth considering.

It is just a little rivulet now, but a good rain storm could turn this into a fence-busting torrent. Special techniques may be advisable to limit future damage.

Another modern option is to use a laser system. The price of construction lasers has come down dramatically in recent years and is a tool that will have plenty of other uses for farm and construction projects. One limitation of construction lasers is that the distance range tends to be only around 300 feet or so, which is plenty for a typical construction site but is a bit short if you're doing a mile of fence. You might have to set intermediate laths for alignment and then continue with another setup from that point. If doing intermediate setups, only go about halfway down the fence, then start again from the other end to reduce the overall alignment error on a long line. If your situation does not permit the use of these alignment tools, an application of geometry can develop sight lines adequate for aligning a row of fence posts, as illustrated in the photo sequence on page 97.

POSTING AT GULLIES

When the fence line goes across a sudden dip in elevation, the posts need special construction methods to keep them securely in the ground. Without extra strengthening, the posts in the bottom of the crossing will tend to be lifted out by the tension of the wire. Depending on how severe the drop in elevation is, one of several methods can be used.

For slight gullies, you can use wire to attach heavy rocks to the gully posts in order to counterbalance uplift forces. If the gully becomes steeper and deeper, you can use oversized posts and set them deeper into the ground. Posts can also be set closer together, adding resistance to uplift forces. For more pronounced gullies, the posts can be secured with bracing to an adjacent post or to a "footer" post placed diagonally in the ground to be highly resistant to uplift.

Fast-moving water can sweep the fence entirely, pile so much debris against it that animals can walk right over, or collapse the soil away so that the fence is left hanging.

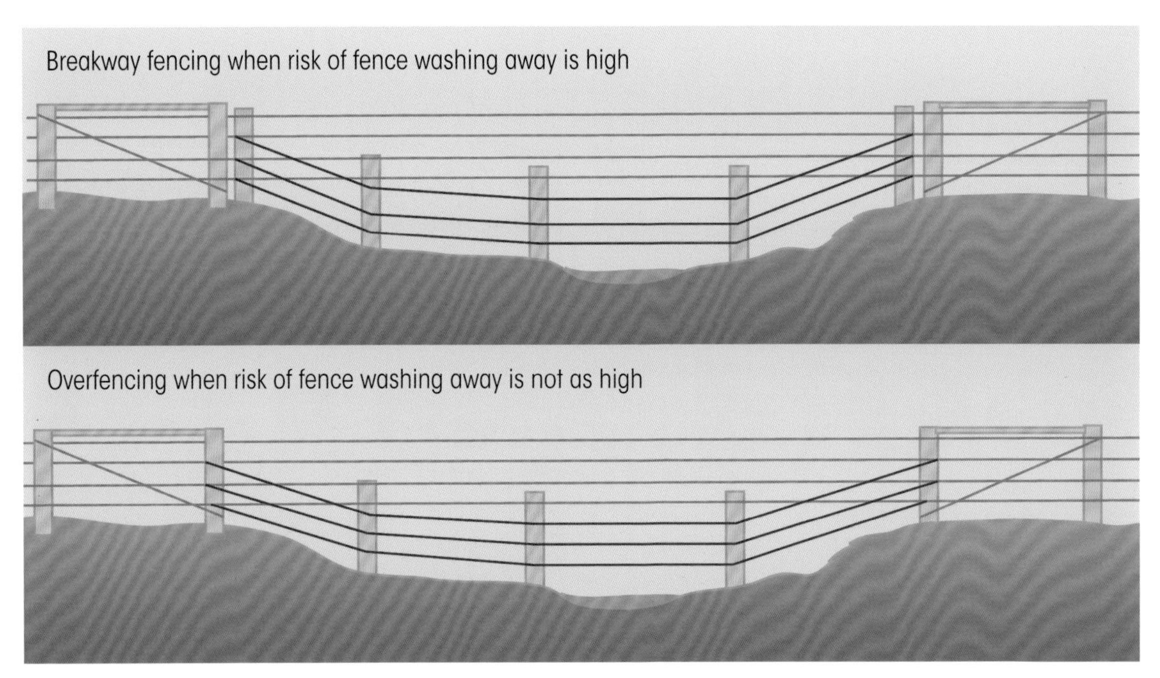

(Above) When a fence crosses gullies, the main wires (gray) can be run right over the gully with either a breakaway section not attached to the main fence (red wires in top diagram) or an underslung section attached at a lower tension to braced start/end posts(red wires in lower diagram). *(Below)* An end post weighted down with a tubful of rocks acts as a counterweight to uplift forces.

CHAPTER

7

INSTALLING WIRE, BOARDS, AND RAILS

This reel of barbed wire is ready for installation. Because barbs catch on each other quite readily, make sure that as you pay it out the reel does not stop turning, which will lead to breaks in the wire.

With the fence posts in place, complete the fence by adding the wire, boards, or other elements developed in your original fencing plan. In general, this part of building the fence will be faster and less laborious, so you can look forward to finishing off the fencing project on a more enjoyable note.

Crimped sleeves form a tight, permanent joint between wires without losing any wire strength.

INSTALLING HIGH-TENSILE WIRE

When you're handling high-tensile wire at any point during construction, take care not to kink it. When stretched, it is very likely to break at any kinked point and will spring back into coils that may strike you.

Firmly attach the wire to the starting post with permanent components and methods of installation. There is no point in making a temporary connection because tensioning the wire will put a lot of strain on the attachment. It is best to have the post attachment completely finished off to its operational form before you go any further.

With the wire tied off at the starting post, you can attach the wire reel to a vehicle and drive forward to pay out the wire. If the terrain is not too rugged, and you'd like to save some time, it is also possible to pay out several high-tensile wires at the same time because they do not tangle up as readily as barbed wire. Using wire from separate spools, attach each wire to the post at the correct height, then slip the spools over a strong, round, steel bar so the spools are side by side. Drive slowly forward to pay out the wires. When you reach the end post, cut the wire and stick the end in the ground to prevent it from recoiling.

A sleeve is crimped on to secure the loop of wire that goes around the post.

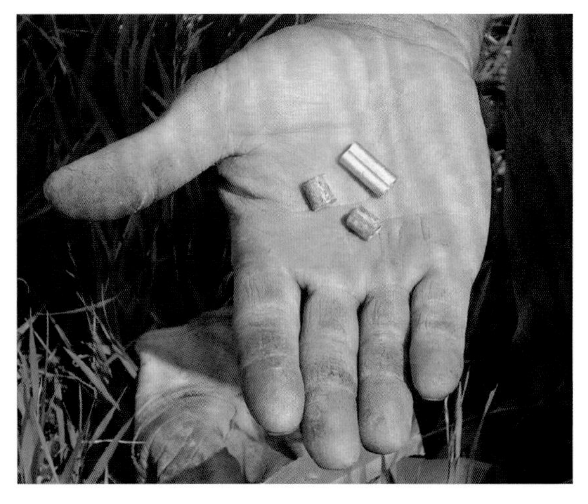

The longer sleeve is for joining wires under full tension, while the shorter sleeves can be used to join the energizer connection wires to the fence wire. The smaller sleeves can be slid on first to be used later.

High-tensile wire is used to make a loop that secures a termination block (white object at center), which insulates other types of electric fencing materials (twine or tape) from contact with wire.

High-tensile wire is stiff, but flexible enough to be tied off if you prefer not to use crimped connectors.

At the end post, attach short lengths of wire tied to the post at one end and to a ratcheting tightener at the other. The wire you payed out from the starting post will be fed into the tightener and wound tight to complete the installation. Ratcheting tighteners are absolutely crucial for maintaining the tension needed for a high-tension fence. While tension springs aren't absolutely necessary, they make the task of setting and maintaining tension so much more straightforward that it is hard to justify not using them.

Stretchers and tension springs can be installed at any point in the fence line, including the starting post. Some guides

A reel of high-tensile wire being readied to pay out behind the tractor for braces and the main line of fence. To act as an axle for the reel of wire, a steel bar has been inserted through the eyes of the tractor's three-point hitch arms.

To begin the diagonal of the brace, four strands of high-tensile wire are strung between posts and the ends of the wire are joined with a crimped sleeve.

A "twitch stick" is inserted between the strands of wire and turned to twist the wires together.

When the diagonal brace is fully tightened, the twitch stick is braced against the top rail to maintain tension.

recommend installation at the middle of the line, while others recommend installation near one of the ends. Having the stretchers and springs relatively close to a gate does make them somewhat more convenient to access for periodic re-tensioning of the fence. The difference in where they are is not nearly as crucial as making sure they are installed somewhere in the line.

With the end posts braced, the main wire is strung out along the new line of the fence.

With the wires tied off at the ends and the stretchers and any springs installed, you can now proceed with attaching the wires to the line posts with either staples or insulated wire holders. Wire holders are best if the fence is to be electrified now or later on.

The tension on wires is much greater on high-tensile fences than on conventional barbed- or woven-wire fencing, which creates a need for staple choice and setting techniques that are somewhat different. These changes will result in stronger, more durable fences and reduced maintenance requirements.

High-tensile fences should use galvanized, 9-gauge, 1¾-inch-long staples. These are only slightly longer than the 1½-inch, 9-gauge staples used for conventional barbed- and woven-wire fences, but the difference is significant. Tests conducted by U.S. Steel show that 1¾-inch staples hammered into wood posts have 50 percent more resistance to being pulled out than 1½-inch staples in the same posts.

Staples on high-tensile fences should never be driven all the way in. Enough

Due to bumping as the reel turns on its axle, the end plate of the reel has come off, leading to a lot of tangled wire. This reel goes back to the shop to be repaired, while another one takes its place.

When high-tensile wire is cut and let go, it will spring back and coil up in inconvenient and potentially dangerous ways. To prevent it getting away like that, stick the cut end into the ground.

Recommended Wire Tension

Once the fence wires are laid out, fencing guides usually mention pulling the wire to certain tension or "tightness" such as 250 pound or 600 pound. In fact, those tension levels are not usually measured by any sort of meter. They could be measured, but it is not generally needed because the range of air temperatures over your day will make the wire tighter or looser anyway.

Use this guide to wire tightness instead.

Wire condition	Tension level	Force of tension
Laying on ground	Loose	0
Pulled up to stapling level	Tight	250 pound (approximately)
Tighter than stapling level	Extra tight	600 pound (approximately)
"Singing tight" or "Banjo tight"	Dangerously tight	Far above 600 pound, so loosen the wire immediately

With all strands laid out, the next steps depend on whether the ground is fairly level or not.

- Level ground: pre-stretch the top wire extra tight, then reduce tension to regular tightness.
- Not-so-level ground: pre-stretch the top wire extra tight, then reduce tension to somewhat less than tight to allow for increases in tension as the wire is pulled upward or downward during final stapling to posts.

of a gap must be left so that the wire can move freely. This freedom to slide allows wire to be tensioned uniformly on long runs. In addition, any local strain on wire will be distributed over the entire fence. Extra strain may come from stock running into or leaning on the fence, or from the wire expanding and contracting due to changes in temperature. Temperature-related changes in tension are why you'll see power line wires with considerable sag in summer; they are strung that way so that they do not snap in winter as the wire becomes cooler and shorter. The principle operates exactly the same way in wire fences, which is why staples are left a bit loose to allow the wire to slide back and forth as temperatures change between winter and summer, or even between noon and night.

Driving staples all the way in increases friction as the wire moves and will result in shorter wire life and more frequent need for repairs and re-tensioning. When stapling, do not use the staple to pull the wire toward a post that's slightly offline. Push the wire against the post before driving in the staple. Before driving, align staple points against the wood grain lines in the post. If the points of the staple are aligned with the grain, they tend to split the wood and reduce the holding power of the staple.

One side of each tip of a staple has a flat "slash cut" in the surface. As the staple is driven into a post, these slash cuts act as wedges and force the legs of the staple to

Ratcheting spools are installed in the fence to allow easy re-tensioning as necessary.

This type of ratcheting tightener uses a crank to turn the spool.

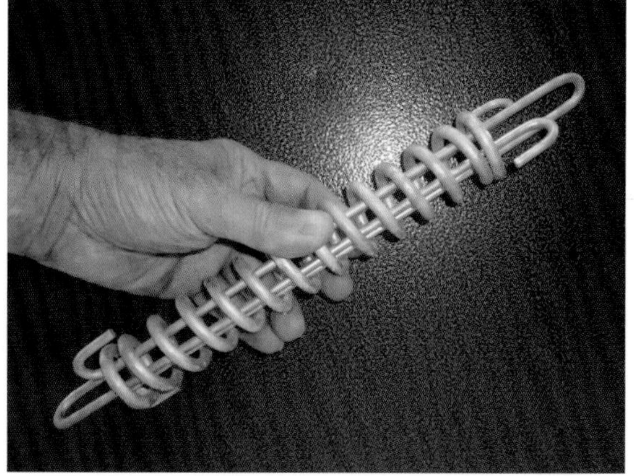

This wire tensioner keeps wire at a specific tension during temperature variations. The spring end is hooked to the wire, and the other end is hooked to a tensioning device, such as a ratchet wheel. As the tensioner is tightened, grooves are exposed, indicating either 150 pounds tension or (if tightened further) 250 pounds tension. As the wire heats or cools, the tension remains at the level set during installation.

curve as they penetrate the wood. Rotating the staple toward the flat side of the slash cut will cause the legs of the staple to converge in the post. Rotating the staple away from the slash cut will cause the legs to curve outward. Staples with legs converging outward will have about 40 percent more resistance to being pulled out than staples with legs curving inward and toward each other.

Posts in low spots will have upward strain from the wires, while posts on high areas will encounter a downward strain. Stapling technique helps counter this extra strain. For posts in low spots, drive the staples in at a slight upward angle. On rises, drive staples into the post at a slight downward angle.

Posts at more pronounced rises and dips benefit from double stapling, which can be done in two ways. The first is to simply double up on the staples and drive two side by side, either tipped upward for posts in dips or tipped downward for posts on a rise. The second way is to drive a staple parallel to the wire so that it acts as a ledge for the wire to push against. On posts in dips, the ledge staple is placed just above the wire and tilted slightly downward so that the wire rests under the ledge and against the post. For posts on rises, the ledge staple is placed just below the wire and tilted upward

Tension may have to be adjusted to allow for dips and rises along the line of the fence.

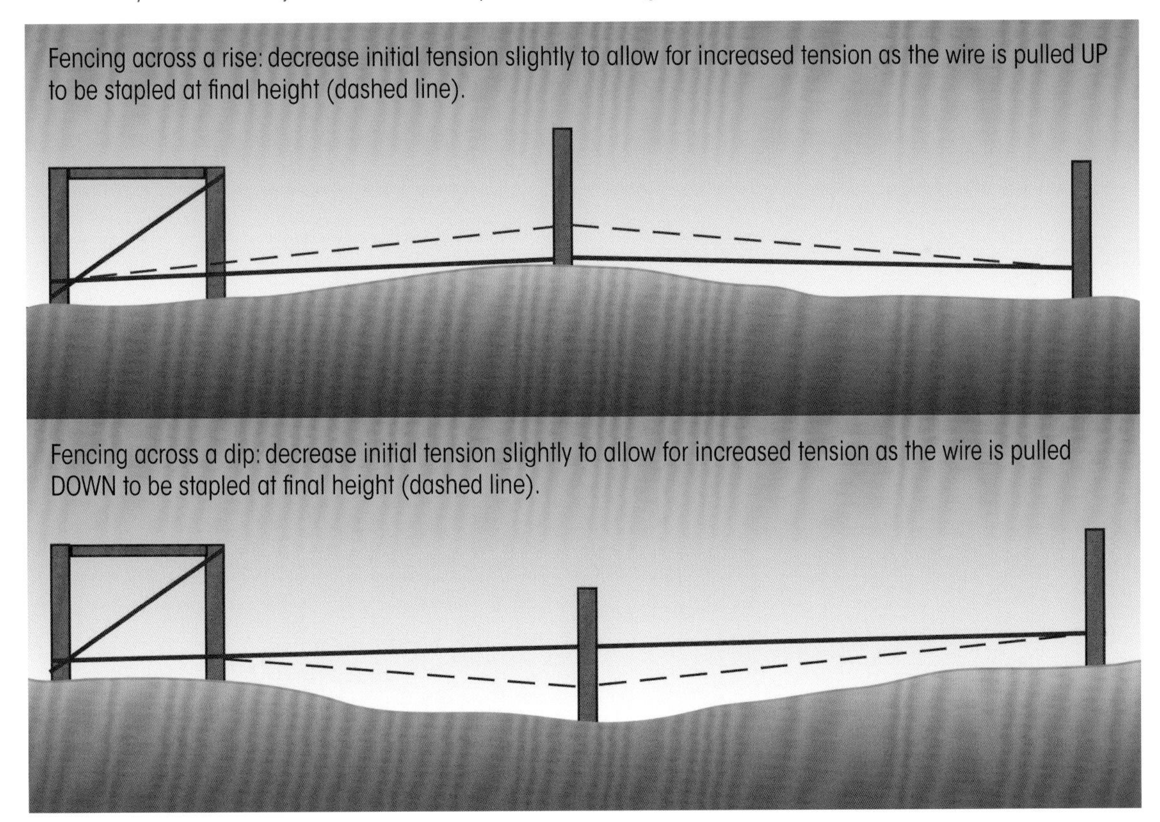

Hitting high-tensile wire while stapling creates weak points in the wire. Instead of stapling the wire to fence posts, special holders can be installed.

Wires clip into the holders. This protects against installation damage and also makes removal easy if the fence needs to be taken down or moved.

Wire clips need to be able to stand years' worth of exposure to damaging UV rays, so choose the best ones you can afford.

slightly so that the wire rests on the ledge and against the post. To finish off this type of stapling, the final staple is driven in as on a regular post. One final caution is that when stapling high-tensile wire, take care not to nick the wire with the hammer. Even a slight nick makes the wire more prone to breaking at that point.

Once the wires are attached to the posts, you can tension the strands using the ratcheting inline tensioners. Start tensioning with the top wire and work downwards. Since each strand, as it is tightened, also puts additional strain on the start/end post, tightening any one strand may slightly loosen the others. For this reason, you may need to check tension on all strands once you've finished tightening the bottom one. There will not be that much change, so final tensioning will be a matter of a little fine-tuning.

Wire Fences and Lightning

When you're constructing any type of metal wire fence that's not electrified, consider the benefits of adding ground rods to protect against collateral damage by lightning. Electric fences are already grounded as part of the wiring, but additional lightning protection should also be added according to the instructions that came with the fence energizer.

If lightning strikes a wire fence, the real danger is not so much to the fence itself (although lightning may blow apart a post that it strikes) but to any living creatures near the fence. The U.S. National Agriculture Safety Database notes that lightning strikes can travel almost two miles along the wires of an ungrounded fence, so the strike does not have to be that close to wreak havoc.

Livestock, especially cattle, will often be near a fence in a storm because they tend to seek shelter by drifting away from the storm and stop when they encounter a fence. When lightning strikes, the easiest path to the ground (if the fence has no ground rods) is right through the animals standing nearby. The animals do not have to be in direct contact with the wires to be injured. Step voltage can radiate out through the ground from a lightning-struck post or tree, which is an effect that results in many livestock deaths every year.

The danger is not confined to livestock. As a storm approaches, you or a family member could be opening a gate in order to bring the stock home to safety. If lightning happens to strike, even up to two miles away, the person near the fence becomes the electricity's path to the ground, again with often-fatal results. For this reason, it may be advisable to place grounding rods within a safe distance (100 feet or so) back from the pivot point of the fence so that a bolt that strikes somewhere down the fence line can reach a safe ground before it reaches a person holding the gate.

Extension services recommend grounding metal wire fences by driving ½- or ¾-inch steel rods or pipes next to the fence posts at least 5 feet into the ground at intervals of no more than 150 feet along the fence. Make sure the ground rod securely contacts all the fence wires. Although corrosion from over the years may increase the electrical resistance in the wire-ground rod connection, the massive jolt in a lightning strike should be enough to overcome the resistance. These slim metal grounding rods are not nearly as hard to drive as a wooden post, so if a post pounder can't be hauled to the location, pounding them in with a sledge or a post slammer is a feasible way to add ground rods to an existing metal wire fence.

If the high-tensile wire fence is to be electrified, make attachments to grounding rods and fence energizers once the wire is completely installed. Current-carrying wire can be attached either to permanent, crimp-on ferules installed during the initial tying off of the wire to start/end posts or by means of removable clamps.

INSTALLING BARBED WIRE

The barbs on barbed-wire fencing are made to inflict pain and will readily catch and tear skin and clothing. Before you start working with this type of wire, be sure to put on protective gear like strong leather gloves, boots, and pants you don't mind getting torn up a bit. There is a reason old-timers called this stuff "the devil's rope"!

Attach the end of the wire to the starting post, then attach the wire reel to a vehicle and drive forward to pay out the wire. Since barbed wire easily tangles up with itself, it is best to roll out one wire at a time. Loosely tie off the wire at the end post and return to the start to string out the next wire. As you string out the various strands, take care to keep them separated. The friction of laying the wire on the ground is usually enough to keep the strands apart as long as you don't put a lot of tension on the strand as you pay it out.

With all strands laid out, the next steps depend on whether the ground is fairly level or not. If it is level ground, pre-stretch the top wire to 600-pound force and reduce tension to the normal 250-pound force. If it is not-so-level ground, pre-stretch the top wire to 600-pound force, then reduce tension to somewhat less than the normal 250-pound

This reel of barbed wire is ready for installation. Because barbs catch on each other quite readily, make sure that as you pay it out the reel does not stop turning, which will lead to breaks in the wire.

The barbed wire here is partially installed and not yet tensioned. While tensioning, you may need to unsnag a strand from the one below it.

force to allow for increases in tension as the wire is pulled upward or downward during final stapling to posts. With the top wire tensioned to its final 250-pound force, fix it in place at the end post. Working downward from the top wire, continue tensioning and tying off wires. Once all the wires are tensioned, staple the wires to the line posts. Staples can be driven completely into the post to help maintain tension on the barbed wire. Be sure to offset the position of the staple points as shown in the photo at right.

Staple points should be offset so they do not pierce the same line in the grain of the post.

INSTALLING WIRE MESH

1. Attach one end of the wire mesh to an end post, standing the wire up to its final position.
2. With the end post now attached, roll out the mesh to the length of the fence. On long spans, splice the ends together as necessary to make the required length.
3. At the end of the mesh, to apply tension to the wire mesh in preparation for stapling, you'll now need to clamp lengths of lumber on both sides of the side of the wire, or use commercially-made stretching bars to do the same thing. Where you place the clamping bars depends on the length of span and whether the ground is fairly level or not.

This wire-mesh splice, showing how the ends of the mesh are overlapped, is cut and twisted back to make the join. This line of mesh could use a bit of re-tensioning.

The wire mesh at the corner shows how mesh is wrapped around the corner post to keep it tightly secured under tension.

If the ground is fairly level and the span is short enough that pulling from one end will achieve sufficient tension:

1. Attach the clamping bars a little ways past the end post. It needs to be far enough past the post that once the mesh is fully tensioned, you can cut the wires one at a time and attach them to the end post.
2. Stand the wire up and lean it against the line posts.
3. Connect one end of a winching device to the clamping bars, and the other end to a suitably immovable object such as a tractor or large tree.
4. Once the mesh is sufficiently tight, cut one horizontal strand at a time and attach it securely to the end post. Continue until all strands are securely attached.

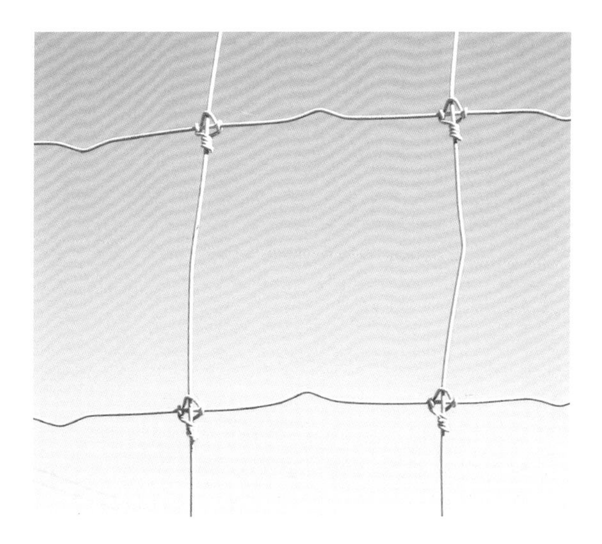

When this wire fence mesh is tensioned during installation, the "kinks" will straighten out to show that sufficient tension has been applied.

If the ground is not so level and the span is long enough that pulling from one end still leaves an unacceptable sag in the middle:

1. Securely tie off the untensioned mesh to the end post, the same as you did at the start post.
2. Pick a spot between two line posts where you can apply tension. For rolling land, this is often at the top of a rise or bottom of a dip.
3. Cut the mesh and attach clamping bars to each end of the mesh.
4. Cut a length of mesh about the same length as the distance between the line posts.
5. Connect one end of a winching device to one clamping bar, and the other end of the winching device to the other clamping bar. It may be handy to have two winching devices in order to apply equal tension force at both top and bottom of the mesh.
6. One at a time, cut each horizontal strand of the tensioned mesh and splice it to the short length of mesh you prepared in step 4.
7. Release the winching devices and remove the clamping bars.
8. Staple the mesh to the other line posts. If it needs to drop down where the land dips, you can step in a lower square of the mesh to force it down. Pulling the mesh up while stapling is not as easy to do while stapling, but it can be made a little easier if you start stapling from the bottom up. You may also be able to insert a lever through one of the mesh squares to help lift the mesh. Square wire mesh (page wire) has a built-in indicator for correct tension. If building the fence with lighter-weight mesh, such as chicken wire, it does not need quite as much tension for successful performance. Just pull it tight enough that it stands up without sagging too much.

This wire-mesh splice, showing how the ends of the mesh are overlapped, is cut and twisted back to make the join. This line of mesh could use a bit of re-tensioning.

(*Above*) A wide selection of electrifiable wires, mesh, and tapes is available to suit any fencing application and budget. (*Below*) Installation is made easier with the same kind of insulating clips used with high-tensile fence wire.

INSTALLING ELECTRIFIED WIRE, TAPE, AND MESH

In almost all cases, attaching nonpermanent electrified wire, tape, rope, or mesh is simply a matter of clipping it into holders built into or stapled to posts. Tensioning is either a matter of hand-pulling to the desired tightness or installing ratcheting or spring-tension devices at relatively low tension compared to permanent wire fences. Refer to product packages for specific instructions related to the wire holders and tensioning devices.

Rubber tubes are another form of wire insulation used when a fence consists of alternating energized and grounded wires.

The ridges on commercially made insulating tubes help keep the tube in place once it is stapled.

An electrified tape gate is used with high-tensile electrified fencing. In use, the tape should be twisted and not flat.

INSTALLING THE ENERGIZER

Make the required electrical connections as outlined in the instruction manual for your fence energizer. The instruction manual will also have details on ground-rod placement and attachment, or you can refer to the general instructions below. Don't cheat on ground-rod placement, because the grounding system is vital to effective operation of the fence.

Drive the full length of each rod into the ground. If that is not possible, place them in trenches where soil will be moist all year. Make the necessary connections to the fence and ground rods using connectors specified in the energizer owner's manual. Follow the energizer manufacturer's recommendations on what type of wire to use to make connections. If the wrong type of wire or connector is used, contact between the terminal and wire could lead to electrolytic corrosion, which will interfere with current flow and reduce the effectiveness of the fence. You should also install an input surge protector and lightning protection.

The fence energizer should be located in a dry enclosure protected with lightning rods.

If you use solar panels, place them away from due south by no more than 20 degrees to the southwest. This way the panel will receive maximum afternoon sun to replace the charge used up in the morning. For year-round use, orient the panel 40 degrees off horizontal. For panels used only during the spring and fall, 25 to 30 degrees off horizontal will be more effective, while winter grazing requires a steeper incline (up to 60 degrees off horizontal) on the panel to catch the lower winter sun.

Detachable lugs allow a secure connection of the energizer leads to the fence wires.

Isolator switches permit areas of the fence to be energized (or not) as needed. They also make troubleshooting much easier when faults occur because the damaged section of fence can more quickly be narrowed down.

INSTALLING FENCE BOARDS

Installing fence boards is generally a straightforward matter of nailing up the boards, but a few tips can make the fence stronger and longer lasting. Make a spacer by cutting a block of scrap lumber the same length as the intended vertical gap between boards. This saves measuring and marking every time you put up a new board.

To hold the board at the right height when nailing, drive a nail partway in at the height of the lower edge of the board. This forms a ledge where the board can rest while you nail it in place. Remove the ledge nail after the board is completely nailed up.

Try to arrange the board lengths so that the ends do not always fall on the same post. Staggering the ends similarly to the way the ends of bricks are staggered in a wall will result in a stronger fence. The first line of boards, such as that at the bottom, could start with a full-length board that spans three or four posts, depending on the spacing. The next course up would start with a shorter board that only spans two posts so that the joint falls on a different post than on the first course. A full-length board can then be used on the second course to maintain the staggering of joints.

Use Ardox (spiral-twist) nails for longer-lasting grip. Ardox nails have a smaller-diameter shank than the same size of smooth nails and are made of better-quality steel. The closer you nail toward the ends of boards, the more chance you have at splitting the wood, which then reduces the holding power of the nail. Try setting the nail a little farther back from the end and driving it at an angle into the post.

When nailing up boards, use washers on the spikes to increase the grip of the spike and help prevent it from pulling through the board.

For constructing pole-type fences, the use of plastic clips speeds up installation and helps prevent horse damage by adding more breakaway capability than you'd get by using spikes or screws.

You can't spell "painting" without "pain," but it does improve the longevity and looks of wooden fences.

SPLIT-RAIL WORM FENCE

The classic snake or worm fence, so named for its zigzag progress across the ground, provides an instant historical look for your farm. This fence is still just as effective as it ever was for confining or excluding horses, cattle, sheep, and hogs, and it provides good cover for birds, small animals, and Civil War re-enactors. Until barbed wire came along, this was a common and practical type of fence, at least where there were plentiful supplies of local wood. Once the hard work of splitting rails was accomplished, this kind of fence could be assembled without tools and without the need to sink postholes. It can be partially or completely disassembled if the fence needs to be moved.

One disadvantage of this fence is that because of the zigzags, it occupies three to four times as much ground as a straight fence. In settled areas where land was becoming expensive or difficult to acquire, this occupied land could be a major economic disadvantage. Additionally, though not too bad for grazing, cultivating all the little triangles of land in the zigzags became more difficult as farm machinery became larger. In response, farmers began to experiment with innovative ideas like stacking rails between pairs of vertical posts. This made a more compact fence but also added considerable labor in digging postholes. Innovation in rail fencing dropped off once cheap, plentiful, reliable barbed wire became available.

The zigzags in a snake fence can take up lots of land.

Split-rail fences need to be made from easy-to-split, rot-resistant wood. American chestnut was a popular choice until chestnut blight problems came along. Large, straight

The log-and-block fence is a variation on the snake fence. Note how the blocks near the ground rest on rot-resistant pieces of railway tie and twisted wire has been added to brace the fence. British Columbia Ministry of Agriculture, Food, and Fisheries

Tips for Building a Split-Rail Worm Fence

- Keep rails off the ground and away from moisture.
- Use flat rocks or short sections of log on the ground where rails cross.
- Use the most rot-resistant wood for bottom rails, since they are the most difficult to replace.
- Extend the ends of the rails about 1 foot beyond the rail they cross.
- Wire the two top rails together at the ends of the fence to make a more solid structure.
- As with any fence, adding a discretely placed strand of electrified wire makes the fence a more effective barrier.

logs without knots, such as the oaks and walnuts that grew in virgin forests, were also widely used. Currently, most split rails are made from cedar. This wood is durable, plentiful, easy to split, and quickly takes on a nice, gray, weathered appearance.

To create a split-rail fence, cut logs to a length of 10 to 12 feet and split down the length of the log. If you're a real stickler for historical accuracy, logs for rail splitting were originally cut four "ax handles" long, which is about 11 feet. Depending on the diameter of the log, 4 to 12 rails can be split from a log. The rails don't need to be dried or seasoned before being used for fencing.

The end product should be roughly rectangular rails measuring about 4×4 or 3×5 inches, or triangular rails equivalent to between 3 and 5 inches in diameter. Between 7,000 and 8,000 rails are needed to build a mile of this type of fence.

The angle of the zigzagging was traditionally set so that the distance

The highly desirable look of the classic split-rail fence is retained in this modern and widely available pre-made variation.

between outside or inside points (similar to wavelength) was 16½ feet (one rod). This made it relatively easier to determine the size of a field by counting the zigzags. One acre equals 160 square rods, so a field 20 rods by 40 rods was 5 acres.

RUSSELL FENCE

The Russell fence is now most associated with the northwestern part of North America, especially the Cariboo-Chilcotin ranching region. This fence makes use of the abundant but relatively small-diameter trees in the region and avoids the need to dig holes for posts. Although strong and inexpensive, especially if using local wood, it takes considerable craftsmanship to build, especially in learning the correct sizes for the wire loops and how to twist the third loop.

(Above) The fence rail has been removed to show the pre-bored sockets that hold the rails. Rails are installed as each post is tamped in place. *(Below)* Russell fence is used in ranch country where wood is plentiful and the ground is too rocky to easily install posts. British Columbia Ministry of Agriculture, Food, and Fisheries

A pair of angled stake poles forms a bipod perpendicular to the line of the fence. The top rail rests on top of the V formed by the stake poles, and the second rail is set in a loop of wire below the V. The remaining rails hang from the second rail in a third loop of wire. Binder or tie poles run from the top of each bipod to the bottom rail at mid-panel. The binder poles are inserted into one side of the top wire loop, then turned to tighten the loop, similar to the way a brace wire is tightened on an ordinary fence post. Once in place, the binder pole ends are wired to the lower rail and form a triangular brace against side-to-side movement along the line of the fence.

(Left) This detail of Russell fence shows how rails and diagonal braces are secured by loops of twisted wire. British Columbia Ministry of Agriculture, Food, and Fisheries *(Below)* For relating triangular split rails to round posts in terms of strength, set one end on the ground and spin it to determine the size of its enclosing circle.

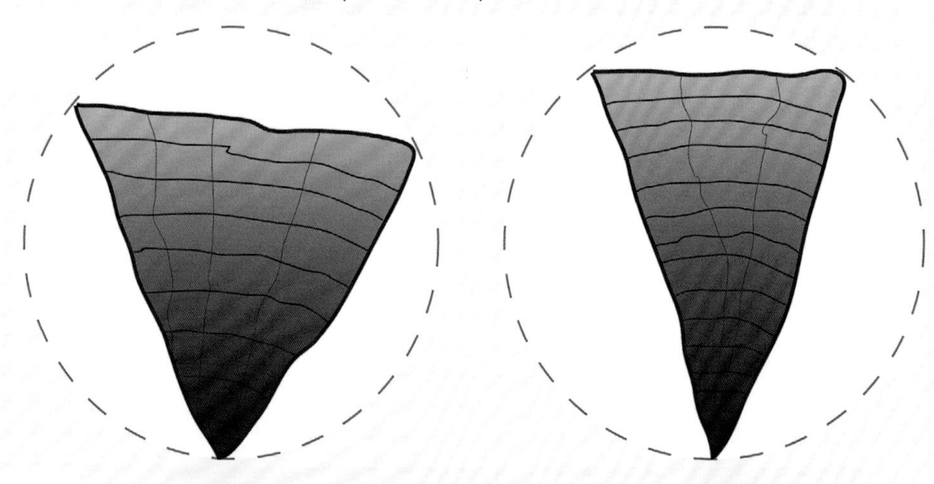

A simpler but less robust type of wood-rail fence employs wooden rails spiked or wired to one side of the bipod supports. Note how rails can be slid aside to make a gate. This type of fence appears in a 1942 photo taken in the Big Hole Valley, Montana, by American Farm Security Administration photographer Russell Lee.

Tips for Building a Russell Fence

Current recommendations for construction of this type of fence include:

RAILS

- Bottom rail 18 inches from ground
- Top rail 50 to 54 inches from ground
- Middle two rails evenly spaced between top and bottom rails
- Panel length (i.e. between stake poles) 12 feet

STAKE POLES

- Minimum 3 to 4 inches in diameter
- Stake pole lower ends must be sufficiently spread to ensure a stable fence.
- In wet soils or areas that are occasionally at risk offlooding, wire a cross pole to the bottom of each stake pole to prevent the fence from sinking.

BINDER POLES

- Minimum 2 to 3 inches in diameter
- Angle binder poles so they cross at the middle of panels.
- The top of each binder pole must lie inside the crossed stakes.
- Attach lower ends of binder poles to the bottom rail with wire.

WIRE

- Number 10 or 12 gauge black annealed wire

If using triangular split rails, see illustration on page 122 for diameter equivalents.

"Wood Fence Construction," British Columbia Ministry of Agriculture, Food and Fisheries, Agdex #724

CHAPTER

8

WHICH FENCE FOR WHICH ANIMAL?

The carpenter created this chicken enclosure with wooden timbers, then lined it with square wire mesh. The enclosure is tall enough to permit humans to enter and clean the chicken yard or to refill the water tank (gray tank at lower left).

Taking into account the experience of generations of farmers, as well as recent innovations that address specific problems, certain types of fence are the usual recommended first choice for each particular situation.

CHICKENS AND OTHER DOMESTIC POULTRY

Compared to four-legged livestock, poultry do not put a great amount of pressure on a fence, so keeping birds within a fixed enclosure is easily handled with ordinary light-gauge galvanized wire mesh woven with octagonal holes (also known as chicken wire). The wire is stapled to the inside of tall posts placed around the enclosure, with stringers between posts providing a place to also firmly attach the top and bottom of the mesh. The post should be tall enough so that the top of the wire is above head height, because you will also need to have wire on top to prevent the birds flying out.

Keeping predators out of the poultry enclosure is a far greater challenge than keeping the birds in. Foxes, coyotes, hawks, and many other predators find poultry an attractive meal. Marauding domestic dogs can go into a killing frenzy in a poultry

This closeup of the chicken pen illustrates how the coarse wire at the base is backed up with narrower-spaced chicken wire at the lower level, where chickens are.

(Above) Doubling up on poultry wire maximizes effectiveness. The finer chicken wire on the inside keeps the birds in, while the stronger and wider-spaced mesh on the outside keeps predators out. Many poultry predators fly in, or climb the fence and jump in. To keep these aerialists out, don't forget to cover the tops of poultry enclosures. *(Below)* Details of portable dog fencing used as a fence for chickens. Standing the fence at angles helps it stand up.

coop, destroying dozens of birds. Other predators such as skunks and magpies are nest raiders, eating eggs while causing stress riots among the birds at the same time.

The mesh placed on top will deter all but the most determined attacks from above by hawks, magpies, crows and other wild birds. It may not, however, be enough to deter animals such as foxes that manage to climb to the top of the enclosure. To stop these predators, you'll need to make sure the edges of the top and side mesh are very securely attached to the top stringers, and perhaps also add a layer of stronger square mesh to the outside of the enclosure. For extra security, a strand or two of electrified wire can be placed on the outside of the enclosure.

Foxes, coyotes, and dogs may also try to get under the wire, so to prevent digging, an apron of wire mesh can be laid on the ground around the outside of the enclosure, then covered with a thin layer of soil to hold it in place. This apron of wire will likely corrode to the point of uselessness within a few years, so plan on checking it for holes every few months and replacing it as needed. Neither the expense nor difficulty of installations is very great.

A new and growing trend in poultry raising, especially on small farms, involves allowing the birds (usually chickens, but sometimes duck, geese, and others) to roam loose in larger areas such as pastures, gardens, or orchards. At night or in bad weather, birds are herded into either a stationary coop or to a portable coop on a skid that can be moved close to where the birds range. Along with getting the birds outside, this allows them to be used for weed and pest control. Chickens are enthusiastic consumers of many types of insects such as grasshoppers. Geese are quite efficient at grazing grasses out from strawberry patches and herb patches, quite a tedious task if done by hand. Ducks will patrol for slugs and snails as well as many insects.

For this type of free-range poultry raising, electric netting is a easy and quick and easy way to install a fence that is quite effective at excluding predators and keeping the birds where they are supposed to be grazing or "working." It will not of course keep the birds from flying out, so you do have to keep their wings properly clipped.

A roomy plywood and mesh chicken pen; in the background is a rabbit hutch.

A detail of the chicken pen, showing how fencing elements are used. The inverted cardboard box provides simple and low-cost shelter.

PIGS

The hard nose disc that makes pigs so efficient at rooting up the soil also makes them very good at rooting under fences to escape. Many of the early fencing efforts in North America were as much about keeping marauding pigs out crops and gardens as they were about keeping other stock in. Wall Street in New York is said to have derived its name from a stout fence erected there long ago to keep free-running hogs from destroying the grain crops planted on the other side.

A strongly built wire mesh fence is the usual solution these days for pig fencing. Electric fencing can also be quite effective because pigs have very little body hair to interfere with electrical contact. Their large, wet noses also make for very good electrical conductivity. Pigs are among the most intelligent of farm animals, so they can quickly learn to understand that they will get an unpleasant shock from an electric fence.

A recommended spacing for an electric hog fence is to have three electrified strands located at 6 inches, 14 inches, and 24 inches above ground. This provides closely spaced wires at intervals up to the height of the snout. Piglets kept with your hogs require wires spaced closer together and lower to the ground. Luckily pigs do not jump, or at least have not yet learned to do so.

For pig fence, a recommended energizer size is one that is able to maintain a minimum of 2,000 volts on the fence line. With fence wires close to the ground as they need to be to contain pigs and especially piglets, a low impedance fence controller is recommended so that it can overcome contact with weeds. A low-impedance fence energizer also helps overcome the coating of dried mud that may insulate the pig from shock.

SHEEP AND GOATS

Fencing approaches for sheep and goats is often treated as similar because of the relatively similar size of the two species and the need for fences to keep predators (mainly coyotes and dogs) out. However, key differences in sheep and goat behavior

Traditional wooden fences actually aren't the best type of enclosure for pigs, as they're very good at rooting underneath. Today the favored solution is a sturdy wire mesh fence or an electric fence.

For a typical small herd of sheep and goats, the main fence has been made with steel tube sections, and wire mesh has been placed inside the main fence.

do make for some key differences in what fences for sheep and goats have to accomplish. In particular, goats are good climbers and seem to be filled with the desire to wander, so fences for goats need to reduce opportunities for individual animals to animal to climb and escape. Sheep very much like to stay in a herd the herd and not generally very bright, so fences need to reduce risk of herd pressure causing the fence to collapse, and also the risk of the sheep getting caught or tangled in the fence and thereby suffering fatal stress.

UPGRADING FENCES FOR SHEEP AND GOATS

For sheep and goat fencing on a small farm, low cost is often a more crucial factor than it is on larger commercial operations. For that reason, when moving to a small farm, owners may find that there are existing fences (e.g., barbed wire cattle fences), and improving them, rather than building something completely new, is often a very attractive option.

Sheep or goats would in most cases easily escape from pastures enclosed by cattle or horse fencing. But by adding mesh or additional strands of barbed or electrified wire, ordinary barbed-wire fences can be improved to the point where they will hold sheep or goats.

Before trying to upgrade an existing fence, make a thorough inspection of whether the existing fence is in good enough condition to be worth upgrading (refer to Chapter 3). There is a risk that if you start off with something in really poor condition (e.g., loose posts, severely rusted wire), you'll quickly need to make a patch once animals are turned into the enclosure. It could worsen to the point where you're burning up too much time and money patching and mending something that was too weak to begin with. By making a thorough assessment beforehand, you can make a reliable decision on whether upgrades will hold until you can afford the time and money to build a new fence.

Barbed-wire fencing for enclosing goats was traditionally thought to be ineffective, but in recent years eight or more strands of closely spaced barbed wire have been found to work well. There is a lot of labor involved with installing and tightening additional strands of wire on an existing cattle fence. But if sturdy corners and line posts are already in place, the cash outlay for wire and staples is quite low, which may be a key factor for many small farms.

When adding additional strands of barbed wire, keep the spacing close together (maximum 3 or 4 inches apart) for the first 2 feet above the ground. Any gaps between the lowest wire and the ground will be quickly found and exploited, especially by goats, so make sure the bottom wires closely follow the terrain. Make sure all wires are well tensioned to prevent development of gaps. Sheep and goats will put their heads through any gaps they find, leading to either escape or entrapment of the animal. The line posts on cattle fence will probably be too far apart to ensure the

Metal panels and wire mesh are used to contain a flock of goats and brown sheep.

consistency of spacing you need for sheep and goats, so use twisted wire battens to maintain wire spacing.

If the end posts are braced with old-style wooden braces, goats may walk up the brace and jump out. Either run the new wires on the inside the braces to deny access to this "goat path" or replace the braces with the more modern type using twisted wire as the triangular brace.

Another type of fence conversion involves adding wire mesh to an existing barbed-wire cattle fence. The lowest strand of barbed wire is moved to ground level, other existing strands are moved to the top of the posts and wire mesh fills the resulting

(Above) Goats are good climbers, so be sure that whatever type of fence you use doesn't allow for easy access to the top. *(Below)* An existing barbed-wire fence with widely spaced wires(top) will be of little use for keeping sheep and goats in and coyotes out. But if the posts are in good condition, adding more strands of barbed wire (bottom) is a way to upgrade the fence to usable condition.

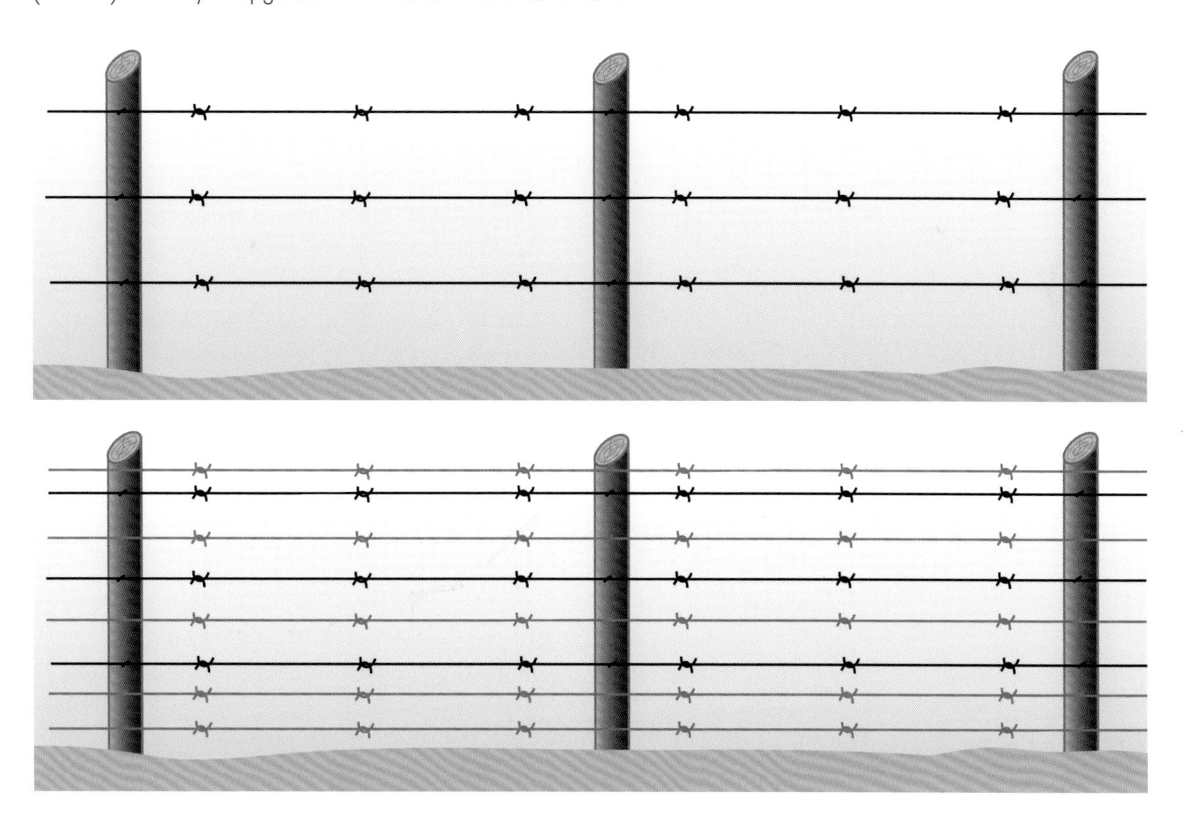

Brace assemblies should be upgraded to the modern type (with a top rail and twisted-wire bracing) to help prevent goat escapes and coyote intrusion.

gap. This type of conversion also involves a lot of labor, but the cash outlay for wire is also relatively modest. The cost for wire mesh is a bit more than for additional strands of barbed wire, but the security is also improved.

When adding mesh, be sure to use the type sold for use in sheep and goat fencing. The spacing between vertical and horizontal wires on this type of mesh is set to not only securely keep animals fenced in, but also to prevent entrapment when animals push their heads into the squares.

When building new, rather than renovating a fence, several modern methods for sheep and goat fencing provide secure enclosure and predator exclusion at reasonable cost. Solutions include:

- For permanent boundary fence, wire mesh with two strands of high-tensile smooth wire above it. Electrified high-tensile smooth wires can also be added at about 7 inches above ground to deter lambs, kids, and coyotes, and about 30 inches above ground to deter adult livestock. Wire aprons can also be installed to prevent digging by coyotes or dogs.
- For shorter-term boundary fence (e.g., on rented land) or separation of pastures on your farm, electrified netting is easy to install and provides reliable containment of stock and exclusion of predators. Requirements for monitoring and maintenance are somewhat higher than for non-electrified wire fences, but sheep and goats require daily monitoring anyway, so checking the fence while you check the sheep is not adding much more labor. As with any livestock, sheep and goats need time to learn that electric fences cause pain and must be avoided.
- For managing rotational grazing, electrified twine or net is a reliable and easily-installed containment and predator

exclusion solution. In this system (also known as control, paddock, or subdivision grazing) sheep and goats are not simply let loose in a large area and left to graze whatever they choose, wiping out the best grass and letting weeds grow unmolested. Instead, they are confined to a smaller paddock within the whole pasture for short periods (with water and shelter supplied) so that they graze everything, including weeds and less palatable plants. After a period of intense grazing plus resultant manuring and "hoof cultivation" of the paddock, they are moved to another paddock, and the one they left is given time to recover and regrow. Within about twenty-one days of decent growing conditions, the resting paddock has fully recovered and regrown, with desirable grasses usually out-competing the weeds and brush. Once a rotational grazing system has been established, animals can be regularly moved from one lush paddock to another, and they seem to look forward to the periodic moves to greener pastures. This system is now widely established as a more effective, environmentally sound method of grazing.

HORSES

Horses or ponies are often the first livestock to be acquired when families move to a new small farm, and can be a major factor in convincing female children to move to the country. Ordinary wire and board fences and corrals can and do provide effective horse fencing, as can be seen on any casual survey of horses and fencing. But certain characteristics of horses make it a wise

Owning a horse is the dream of many a young child, and perhaps for this reason it's often the first animal on a family's new farm.

Barbed wire will hold horses, but for many reasons it's far from the best kind of fence for this application.

choice to install safer and more effective long-term fencing as soon as you the time and funds to install it.

Visibility is a key point in fences for horses, because they can have a hard time seeing a fence. Compared to other livestock such as cattle or sheep, horses' eyes are located more to the sides of their heads. This provides horses with excellent peripheral vision, but makes it harder for them to see small things directly in front of them, especially while running. For a horse, the wires in ordinary wire fences are hard to see, so they may easily crash into ordinary wire fences.

Note how this typical cattle fence is actually a bit low for horses. If you are turning horses into an old cattle pasture, raise the top of the fence and flag the wires so the horses can easily see it.

The flagging on this fence illustrates how it used to help horses avoid blundering into it. Horses are generally unable to see fence wire alone.

Even small strips of cloth tied to the fence wires make it easier for horses to see and therefore avoid a fence.

Wide plastic tape installed on the fence makes it very visible to horses. Installing it near the top makes the height apparent to help prevent jumping attempts.

This is a closer look at the special holders used for solid visibility tape. The holders improve the service life of the tape by avoiding the potential for rips and cracks that would occur if the tape were pierced with nails or screws.

Different Types of Horse Fences

Requirement: Electrified temporary horse fence
Solutions:

- One or two strands of electrifiable tape or polypropylene (PP) rope. Use tape or rope that has interwoven contrasting colors to make it highly visible to horses.
- Easy to install or remove; hand-tighten between posts.
- Corner/end posts can be small-diameter, hand-driven steel or wood posts.
- Line posts can be spaced up to 25 feet apart and are usually the plastic step-in type.
- Small-diameter fiberglass or metal posts create a spearing hazard if a rearing horse descends on them. For metal T-posts, plastic top caps are available to reduce spearing hazards.

Requirement: Electrified temporary, but longer term (up to five years' use), horse fence
Solutions:

- One or two strands of electrifiable polyethylene (PE) rope. Use tape or rope that has interwoven contrasting colors to make it highly visible to horses.
- Easy to install or remove; hand-tighten between posts.
- PE rope provides greater resistance than PP rope to deterioration in sunlight.
- Corner/end posts can be small-diameter, hand-driven steel or wood posts.
- Line posts can be spaced up to 35 feet apart and may be plastic step-in type. For longer-term installations, use wood or insulated-steel posts.
- Small-diameter fiberglass or metal posts create a spearing hazard if a rearing horse descends on them. For metal T-posts, plastic top caps are available to reduce spearing hazards.

Requirement: Lower-cost electrified permanent horse fence
Solutions:

- Up to five strands of electrifiable polyethylene (PE) or polystyrene (PS) rope. Use tape or rope that has interwoven contrasting colors to make it highly visible to horses.
- For humid climates, PE rope retains attractiveness longer because it is more resistant to mildew damage. For dry climates, PS rope is longer lasting under intense sunlight. Strength and conductivity of PS and PE rope are similar.
- Corner/end posts should be power-driven or dug-and-tamped wood posts.
- Line posts can be spaced up to 35 feet apart; steel T-posts or wood. For metal T-posts, plastic top caps are available to reduce spearing hazards.

Requirement: Higher-visibility electrified horse fence
Solutions:

- Up to five ribbons of wide electrifiable tape. Use tape or rope that has interwoven contrasting colors to make it highly visible to horses. Tape with interwoven nylon strands is more resistant to sagging.
- More visibility than rope fences but involves more work and cost to install.
- For boundary fences, four ribbons are usually sufficient, but a fifth may be needed for very active or aggressive horses. For inter-paddock fences, a single ribbon may be sufficient.
- Corner/end posts should be power-driven or dug-and-tamped wood posts.
- Line posts can be steel T posts or wood and spaced 15 to 20 feet apart. For metal T-posts, plastic top caps are available to reduce spearing hazards.
- Twist tapes two to three times between posts to keep tape from flapping in the wind. Flapping makes the tape wear out and break more rapidly and may stress out the nearby horses.

Requirement: Permanent electrified horse fence with better visibility than rope, less cost than all-tape.
Solutions:

- Combination fence with three to four strands of electrifiable rope and an additional ribbon of electrifiable tape at the top of the fence.
- Adding an upper wide tape makes the fence very visible to horses, while the lower ropes provide a visually discreet, lower cost, and very effective barrier.
- Installation tips for rope and tape as described in previous sections.

Requirement: Permanent non-electrified horse fence
Solutions:

- Wire mesh: Use a close-spaced mesh to keep the horses from damaging themselves on the wire and to help keep children, dogs, and other animals from easily climbing the fence.
- Plastic monofilament line: This type of fence is larger, more visible, smoother, and more elastic than high-tensile wire and is therefore safe enough for use as horse fence.
- Wood or vinyl board fences, rail fences.

With any of the above non-electrified horse fencing solutions, adding a strand of electrified wire 2 to 6 inches inside the fence helps keep horses from fighting across the fence or cribbing, chewing, rubbing on, or pushing against the fence.

(Above) This horse fence shows the use of wide tape for electric fencing for horses, and how the top of the fence has been raised to horse height. Even though it was a hot day, this horse is wearing a "coat" to protect it from bugs. *(Below)* Bored or nervous horses will chew on wooden fence boards, which can lead to various ailments.

Horses' coats are also more easily torn than other livestock, and there is much greater aesthetic objection to having a skin injury to a horse than there is to ordinary livestock. So ordinary barbed wire or high-tensile wire, both of which can easily damage horses' coats, are not the best solution.

Horses also suffer more obviously from boredom, which can lead to nervous habits such as cribbing (gripping a fence board in their teeth, arching the neck, and pulling backwards while swallowing air and grunting) or actually gnawing on wooden fences. Either of these obsessive-compulsive behaviors can be very damaging to both the horse and the fence.

CATTLE

The main issue involved with fences for cattle is standing up to the tremendous amount of pushing and rubbing that these large, heavy beasts apply to fences.

Barbed wire (see pages 55-56) is the classic solution for permanent cattle fence. One or two strands of barbed wire will often easily contain cattle, although four to five strands are more common in well-built fences. The disadvantages of barbed wire include relatively high cost and the need for frequent maintenance, i.e. tightening wires that have sagged due to pressure from animals or action of the weather.

High-tensile wire fencing for permanent cattle fence is rapidly taking over from barbed wire due to lower costs, easier installation, and reduced requirement for maintenance. High-tensile wire fences can also be electrified if desired.

If wood, labor, and time are abundant and you'd like a historic look to your farm, rail or Russell fences are effective permanent cattle fences. They also avoid the work of having to set posts, an important consideration where soils are stony or very hard.

Temporary cattle fences are also required, especially for managing the rotational grazing now widely established as a more effective, environmentally sound way to graze cattle. In this system (also known as control, paddock, or subdivision grazing) cattle are not simply let loose in a large area and left to graze whatever they choose, wiping out the best grass and letting weeds grow unmolested. Instead, they are confined to a smaller paddock within the whole pasture for short periods (with water and shelter supplied) so that they graze everything, including weeds and less palatable plants.

After a period of intense grazing plus resultant manuring and "hoof cultivation" of the paddock, they are moved to another paddock, and the one they left is given time to recover and regrow. This short but intensive type of grazing is incidentally thought to be similar to the way native bison grazed the plains in days gone by. Within about twenty-one days of decent growing conditions, the resting paddock has fully

When you're planning a fence for cattle, remember that it must be strong enough to withstand enormous amounts of pressure from the animals pushing and rubbing against it.

If your barbed-wire fence has sagged, due to weather or animals, or just age, cattle can easily find their way out.

recovered and regrown, with desirable grasses usually out-competing the weeds and brush.

Once a rotational grazing system has been established, cattle can be regularly moved from one lush paddock to another, and they seem to look forward to the periodic moves to greener pastures. But for rotational grazing to work, an effective system of temporary fencing is needed to keep cattle from wandering away from where their grazing activity needs to be concentrated.

You could, of course, establish permanent paddocks with barbed-wire or high-tensile wire fence. But besides being very expensive and maintenance-intensive, that would not allow you to alter and improve the size and composition of paddocks.

That's why the preferred solution for paddock management is now temporary electric fences. They are inexpensive, quick to install, and very effective. For cattle that have already learned to respect electric fencing, a single strand can be enough to keep them where they need to be, and away from such delights as stored hay or a cooling stream. For calves or cow-calf pairs, more strands will be needed until a few touches to the fence train the young cattle to respect it.

Electrified tape, rope, or twine can be used. Of the three types, tape is the most expensive, most visible, needs more line posts, and takes the most work to install. Twine is the least on all counts, with rope falling in between. The reduced visibility for rope and twine is not as much an issue on managing cattle movement is it is for damage from deer in the area.

Another method of cattle "fencing" that's still not fully researched is the use of visual barriers that take advantage of the way cattle see. Cattle are quite reluctant to cross areas of alternating light and dark, which is one the reasons why grate-type cattle guards are effective. Cattle graziers have recently been experimenting with using simple wooden grates placed, for example, at stream banks to serve as a barrier to cattle movement. The advantage of the grate, rather than a conventional fence, is that is does not generally impede the movement of other types of animals using the area.

The line of yellow posts taking off from this barbed-wire boundary fence marks the installation of an electrified fence to manage rotational grazing.

BISON

Bison (buffalo) meat is becoming more and more popular in North America for many reasons, not least the taste and reduced fat content compared to beef. As a result, bison ranchers are in the midst of a lot of research and trials of production techniques for farm-raised bison, including how to provide effective fencing. Bison are large and powerful animals that can, if they want to, smash their way through almost any fence with dismaying ease.

Strangely enough though, they seem to accept and respect even a relatively weak wire fence once they have come to believe that it marks the boundary of the territory where they have good food, good water, and good treatment. Most experienced bison raisers are said to keep herds tame and manageable by taking the approach of leading the herd, not driving it. Bison are treated to special goodies such as range cake or molasses seem to become quite agreeable to staying where the goodies are. As one bison producers' web site (nwbison.org) puts it "Bison have been observed staying put in a pasture with a rickety fence that wouldn't hold a crippled cow."

Fencing approaches for bison ranching currently varies from upgraded cattle barbed-wire fences to 6-foot, high-tensile page wire to various types of electrified fence. A common type of fence simply

If bison really wanted to get out, nothing would stop them. But if they like where they are, almost any decent fence appears to be enough. As with any livestock, having a well-built fence helps show you've been diligent about keeping your animals away from collisions with vehicles.

upgrades cattle fence by adding several extra strands of barbed wire. The end result is a fence with five or six lines of wire spaced 8 to 10 inches apart for a completed height of 5 feet.

If bison are being raised close to a road or populated area, it's probably a very good idea to put extra materials and effort into building a sturdy fence. If these animals do get out and get into damage such as a collision with a car, the consequences will be expensive for you, the owner of the animals. Having a sturdy fence in place helps show you have taken proper care to prevent escapes and their aftermath.

A Saskatchewan extension bulletin notes that fences constructed using multiple lines of smooth high-tensile wire are gaining popularity due to their lower cost and ease of construction. Typical smooth high-tensile wire fences are constructed with six or seven lines of wire spaced 8 to 10 inches apart for a completed height of 5 to 6 feet. This type of fence construction offers the advantage of easily electrifying several lines to control problem livestock.

The same bulletin goes on to note however that some bison producers are turning away from high tensile wire and going back to low-tensile barbed wire, even though the barbs may have little effect on the thick, matted coats of bison. The apparent problem with high-tensile wire is that when it breaks, the wire will attempt to spring back into the shape of the original spool it came on. This will result in a tangle of wire that may snare and cripple bison.

DOMESTIC DOGS

It would seem logical and desirable that once you move to a farm, your dog or dogs can run free, as they traditionally did on farms. That's not usually a problem if you're always there, or take the dog with, as farmers traditionally did.

But it becomes a huge problem in areas with many small farms where the owners have off-farm business during the day and the farm dogs are left at home alone. These normally well-behaved, lovable dogs very often get into running with a pack with other bored dogs, and the pack behavior quickly reverts to something resembling the dogs' wild ancestors. Mass kills of chickens, mutilation of sheep and running

Dogs very quickly learn that electric fences must be avoided and will continue to do so even when the power is off, as in this photo.

Electric fencing for dogs is definitely worth considering, especially when the alternative may be to construct a high-security, dig-proof, climb-proof kennel area such as this.

to death of cattle and horses can be the result. Unlike wild predators, domestic dogs don't just kill enough to eat–they continue to chase and kill as long as the opportunity presents itself.

It's all very difficult to deal with when the dog pack members, knowing their owners always come home at a certain time, run back home to wait innocently on the step. Alternatively, if "caught in the act," domestic dogs may quite legally be shot by farmers defending their stock from depredation. Either way, some very bad feelings have been seen to erupt between neighbors over this problem. Farmers are angered and horrified by dogs killing or chasing stock, while dog owners are equally angered and horrified that anyone would harm their precious pet.

Dogs out running in the fields and woods are also quite vulnerable to predator attack. A well-known trick of coyotes is to take turns at tiring dogs out by luring them farther and farther away from home. Once the dog is exhausted and well out of sight of its human protectors, it becomes an easy task for the coyote pack to surround, kill, and eat your dog.

A territory fence for the dog helps prevent most of these situations. Dogs do actually seem to like staying in and guarding what they have come to view as their territory, so providing a fence to keep them home is not so much an imposition on their freedom as it is giving them a job they like to do. The fenced area can be big enough to give the dog plenty of room to move around without the restriction of being tied up.

Wire mesh 3 feet to 4 feet high is an inexpensive and easy-to-install fencing solution for dogs. Wire mesh can also be attached to the back side of more attractive wooden yard fences, such as the split rail fence shown on page 69. One method that is sometimes overlooked these days is the old time picket fence. The fence provides secure containment while the spaces between slats allow the dog to keep an eye on what's going on outside the fence.

Some dogs do however learn to be expert fence climbers, and most breeds, especially terriers, can make a determined effort at digging under the fence. Electric fencing then becomes a good solution because it keeps a dog from even approaching the fence. Like other animals, dogs need to be trained to understand that they will receive a shock when trying to breach the fence. A good way to do this is to put bait such as peanut butter or empty salmon tins on the wire. This prompts the animal to touch its nose or tongue to the wire, providing a very good shock with no lasting damage.

Electric mesh, similar to that used for other small animals such as sheep and poultry, is also an effective, relatively inexpensive, easy-to-install and visually non-intrusive fence system for dogs.

EXCLUDING WILD ANIMALS

Along with fencing in your gardens and farm animals, fences can also do much to exclude wild animals that might consider farm gardens and animals quite a tasty snack.

DEER

Seeing deer is often quite a thrill when going to the country, but many rural residents are finding that deer are widely regarded as one of the worst pest problems in crops, gardens, orchards, and vineyards. Natural predators of deer, such as wolves and mountain lions, are no longer present in many areas, and where other checks on population are no longer in place, deer population levels can boom, leading to great vexation among landowners as deer seek food.

Building fences good enough to keep deer out is quite challenging, because deer can easily jump fences less than 8 feet high and are also quite adept at crawling under fences. They may also crash right through a fence when they fleeing from real or perceived danger, or if they simply don't see the fence while bounding along.

Wire mesh fencing at least 8 feet tall can provide an effective barrier with relatively low maintenance requirements, but tends to be a more expensive option. A typical 8-foot high vertical wire mesh fence can be constructed from two 4-foot high sections of mesh joined with hog rings (C-shaped wire clips that are pinched together) or loops of ordinary wire. Two or more strands of barbed wire, spaced 10 inches apart can be added to the top to extending the overall height.

High-tensile electric fencing can be a suitable deer fence where deer pressure is not as intense. High-tensile electric fence is less costly and easier to erect but does need more monitoring and maintenance.

Deer can easily jump fences less than 8 feet high and can also crash through a fence when running. Wire mesh and high-tensile electric fences over 8 feet high can be effective barriers.

Deer also need to be trained to understand that they will receive a shock if they approach the fence. Attractants such as special scents or peanut butter can be put on the electrified wire. This prompts the animal to touch its nose or tongue to the wire, providing a very good training shock with no lasting damage to the animal.

Tall electrified plastic mesh can provide a good barrier, especially if you can back it up by equipping the enclosed area with one or more large dogs. The mesh is quite effective at containing dogs, and deer see the dogs as predators.

Taking advantage of certain aspects of deer behavior can improve fence effectiveness.

- Deer are hesitant to jump a wide barrier. By slanting a fence toward the deer, or adding a second visible strand of electrified wire about 3 feet away from the first, the fence starts looking too wide for the deer to safely attempt a jump.
- If the ground slopes upward toward a fence, it looks taller to deer and makes them more hesitant to jump. Conversely if the ground slopes downward toward the fence, it looks lower and deer will be more confident about jumping. Where the ground slopes down toward the fence, add additional height to the fence or, if possible, relocate it away from the slope.
- If a new fence crosses existing deer trails, deer will try hard to continue traveling that trail. The fence must be easily seen and effectively constructed at that point to break their habit of going there.
- Deer generally prefer to stay near areas with good cover. Establishing open cleared areas around deer food sources will make deer more hesitant about exposing themselves to view while crossing open ground. Do not rely too much on open areas for deer movement control, because deer often simply wait until dusk or predawn so that they slip across open areas unobserved.

BLACK BEARS

In rural areas, the common black bear can be a troublesome raider of crops, gardens, orchards, and garbage. Black bears are now found in forty-three states, are abundant or common in twenty-nine of them, and have recently reoccupied much of their original range in the United States, especially in the east. During the American colonial period and after, black bears had been hunted almost to extinction in populated areas, while logging and clearing of land for farms reduced bear habitat. But as small farms failed and people moved back to the cities, bear habitat slowly recovered and populations started to increase.

With once-abandoned small farms now being reoccupied, or large tracts of woodsy land being subdivided for new small farms, bear problems are becoming more frequent.

Black bears are quite common in rural areas, especially in the eastern United States. A sturdy electric fence is the best way to keep them away from your farm.

Many food sources on small farms can attract bears, and once they have discovered a good feeding site, they are diligent about coming back.

Electric fencing is an effective, easy-to-install, and relatively inexpensive way to prevent black bears from entering. Since bears are large, powerful animals that will quickly exploit any weakness in the fence, do not buy cheap materials to reduce costs. Install electric fence early in the season so that a bear receives a strong, negative experience the first time it attempts to access the area. It is much easier to keep bears away from an area before they have already learned there is good eating to be had there.

Bears, like other animals, need to be trained to understand that they will receive a shock when trying to breach the fence. A good way to do this is to put bait such as peanut butter or empty salmon tins on the wire. This prompts the animal to touch its nose or tongue to the wire, providing a very good shock with no lasting damage.

Recommended electric fencing to exclude bears:

- Minimum six strands of 12.5-gauge high-tensile galvanized wire tightened to a minimum of 125 pound tension at 72 degrees F. High-tensile wire ensures that the fur of the bear is parted and the wire touches the skin directly. Use suitable insulators on all posts.
- Install posts no more than 25 feet apart, set into the ground at least 2 feet. To maximize strength, posts are best pounded in rather than tamped into pre-dug holes.
- Install the bottom wire approximately 2 inches above the ground, to prevent bears from trying to squeeze under the fence. Fence line clearing and/or closer spacing of posts may be required to keep the bottom strand this close to the ground.
- Install the top wire approximately 44 inches from the ground, and space remaining wires spaced evenly in between. Use suitable insulators on all wires.
- Fence effectiveness can be improved by adding an additional "stand off" strand of charged wire installed 12 to 18 inches out from the fence and 8 to 12 inches above the ground.
- In areas with loose or sandy soil, install a horizontal wire apron to keep bears from digging under the fence. A 6-foot width of chain link or page-wire fence should be attached to the fence posts, laid flat and secured to the ground at the base of the fence. Covering the apron with an inch or so of soil, or weighting it down with plenty of rocks keeps the bears from lifting it up and digging under.
- Use alternating hot and ground wires to ensure that the bear still receives a shock even if there may be poor conductivity between the bear's feet and the ground due to leaves, branches etc. Connect the first (bottom), third, and fifth strands to the negative terminal on the charger, and the second, fourth, and sixth (top) strands to the positive (hot) terminal on the charger.
- Provide system grounding by connecting the negative terminal of the fence charger to three 5/8-inch ground rods, each driven 6 to 8 feet deep and spaced at least 10 feet apart. Where possible, locate the ground rods in moist soil to improve conductivity. Try to use longer ground rods on very dry sites. On sites with exposed bedrock, use ground mats of wire mesh.
- Choose an energizer of one joule or larger, capable of delivering a minimum shock of 6,000 volts. Secure the energizer in a weatherproof enclosure to prevent damage by weather or bears.

- Install a lightning diverter to channel lightning strikes into the ground rods to prevent damage to the energizer.
- Electric gates (e.g., using strands of electrifiable tape) can also be installed where required.
- To prevent vegetation from growing up beneath the fence and touching the strands, which can drain power, partially bury landscaping cloth along the fence line prior to post installation, or apply non-selective herbicide or soil sterilizer along the fence line.
- Post warning signs to identify the site as electrified.
- Check the fence regularly during bear season. Look for adequate fence voltage output, fence damage from fallen trees, deer strikes, etc., adequate wire tension, any indications of animals trying to dig under the fence, and vegetation that may short out the fence.

RACCOONS, OPOSSUMS, SKUNKS, BADGERS, WEASELS, AND OTHER SMALL WILDLIFE

The gardens, crop field, orchards and vineyards, crops, and lawns on small farms are quite attractive feeding sites for many types of small four-footed wildlife. Since these animals are secretive and most often intrude at dusk or dawn when they are hard to see, it's difficult to drive them off, and some type of fence is needed to maintain a constant, reliable barrier. Scare tactics such as noisemakers or flashing lights may work for a while, but once the animals observe that no physical harm results, they tend to ignore what was scary at first.

These kinds of small animals have characteristics that make ordinary fencing difficult. They are often small enough to squeeze through wires or mesh, and may also be accomplished diggers or climbers, enabling them to get under or over the fence. The light body weight and small footprint of any small four-legged pests makes them harder to shock with electric fencing. Keeping electric fence low enough to the ground to influence these small animals also puts the fence down at heights where weeds and grasses can short out the low wires.

Using a closely spaced mesh and higher voltages can overcome many of these problems and keep small four-legged pests out. Adding one or two electrified strands to a barrier fence is often a very effective combination. Once animals learn that receive a very unpleasant (but harmless) shock from the electrified wire, this denies a place to stand while attempting to dig under, climb over, or tear through physical barrier. Animals may learn to respect an electrified fence wire by touching their

Raccoons are clever, persistent, and acrobatic. A combination of wire mesh fencing (with a buried apron) and a single strand of electrified wire is most effective at keeping them away from your gardens and ponds.

nose to it as they first explore the fence. An even better way to quickly train them is by getting a very effective jolt from touching their tongues to the wire. Placing peanut butter or other attractive bait on the wire encourages animals to lick the wire–once.

With electrified fences, it may appear that since many of these pests are only active from dusk until dawn, the fence energizer only needs to be on for those hours. But don't overlook the capacity for a hungry, determined, or curious animal to test the fence at "off peak" hours. Once they learn that there are times when the electrified fence does not shock them, they may crash through even when the power is on and they do receive a shock.

If pest pressure is intense, providing a sacrificial food plot some distance away can also help manage their movements. If there's a feeding site that's easier to access than the site you wish to protect, pests usually prefer to take that easier and more convenient option for feeding.

OPOSSUMS

These tree-dwelling animals will mainly test attempt to climb the fence. Discourage this by using 4-foot-tall poultry mesh having the top 12 to 18 inches of the mesh bent outward and not attached to any support. Leaving the top of the fence free to bend under the weight of the animal makes the opossum less sure of its footing and more reluctant to climb the fence.

A wire fence can be made more opossum-proof by stretching a "stand off" electrified strand 3 inches out from the wire near the base of the fence. This denies the opossum a place to stand while starting its climb. If they jump onto the fence from nearby trees, the base of the tree can be electrified by winding a strand of electrified twine around the trunk.

RACCOONS

These clever and dexterous animals will not only dig under or climb over fences, but will also patrol for weak points in the fence and gain access by ripping off loose boards or enlarging holes in wire mesh. Raccoons are enthusiastic consumers of sweet corn, melons, and fruit, and will also roll back newly laid sod searching for insects and grubs. They are also accomplished at catching ornamental or game fish in ponds.

To keep raccoons out, permanent wire mesh fences should be strengthened and equipped with a buried apron of wire mesh outside the fence to prevent digging. The fence can then be improved by adding a single strand of electrified strand of wire 8 inches above the ground, standing off about 8 inches out from the base of the fence.

A low two-strand temporary electric fence can be very effective at excluding raccoons from crops, ponds, and areas with freshly laid sod. The recommended spacing is to have one wire 6 inches above ground and the second 12 inches above ground. T-posts or step-in posts can be used to simplify installation and removal of the fence.

SKUNKS, BADGERS, RABBITS, WOODCHUCKS, AND OTHER LARGER DIGGING ANIMALS

Use raccoon fencing recommendations.

SQUIRRELS, BUNNIES, CHIPMUNKS, AND OTHER VERY SMALL ANIMALS

Use opossum fence recommendations, with the addition of a wire mesh size small enough to block these small-bodied animals. If only these small animals and not opossums are the problem, the fence height of the barrier fence could be reduced to 18 inches to make human access easier. Adding strands of electrified wire 4 and 8 inches above ground will likely be necessary to make the barrier fence effective.

WATERFOWL

Wild geese and to a lesser extent wild ducks can be very annoying pests of ponds, lawns, and gardens. Canada geese not only eat and trample gardens and crops, but also leave large amounts messy guano all over. Geese with goslings also become quite aggressive about defending what they come to regard as "their" territory, and may chase you or your children out of your own garden. The type of fence needed depends on observation of whether the waterfowl are mainly flying in or walking in. Wild ducks, for example, mainly fly in while Canada geese more commonly walk in from areas where they are grazing.

To stop ducks and geese from landing, raised grids can be constructed of thin cable visible to both humans and waterfowl. Because light cable is used, several hundred feet can be supported between ordinary steel T posts or equivalent.

Grids on 20-foot centers will stop geese, and grids on 10-foot centers will stop most ducks. If you're installing grids in areas where people and equipment need access, make sure to install the grids somewhat higher so that access is not hindered.

When installing grids, attach each line separately so that if there is a break, only one line falls down, not the entire grid. Wind will make cables chafe against each other where they cross, so eliminate that by zip-tying or taping the cables together at crossing points.

To exclude geese walking in from their grazing area, install a 30- to 36-inch tall fence at the edge of ponds, pastures, crops, or gardens. Poultry wire or electrified netting is generally effective. Where wire fencing would detract too much from the look of the area, fences made of monofilament fishing line have also proven effective. This type of fence is made using 20-pound test or greater line strung 6 inches off the ground, then at 6-inch intervals to the desired height. Thin strips of aluminum foil at intervals of 3 to 6 feet along the fence line help reduce accidental breakthroughs by increasing visibility to people and other livestock.

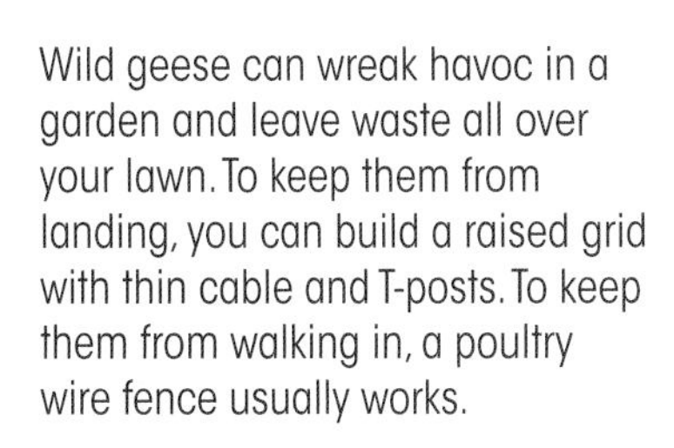

Wild geese can wreak havoc in a garden and leave waste all over your lawn. To keep them from landing, you can build a raised grid with thin cable and T-posts. To keep them from walking in, a poultry wire fence usually works.

CHAPTER

9

FENCES FOR ENVIRONMENTAL MANAGEMENT

A simple wind barrier fence goes a long way toward improving the environment for livestock. In the case of cattle, it also helps keep them from trying to escape storms by walking downwind.

Discussion of fence types usually focuses on managing the movement of farm animals and their predators and, to a much lesser extent, the movements of people, such as recreational users. But there are also certain types of fences used for environmental purposes, such as managing wind, water, and snow. These types of fences can help prevent soil erosion, trap moisture for crops or livestock, increase road safety, and make farmyard living conditions more pleasant.

Improving your own farm microclimate is a very important benefit that can be achieved with rows of bushes and trees that form a living fence or shelter belt.

In many cases, shelter-belt trees or bushes are planted with spaces in between for ease of weed control by cultivation during establishment. But in dense plantings to form a living fence, the bushes themselves can outcompete weeds by shading them out, so once the fence is established, cultivation is no longer required. Spreading a thick, woody mulch on the soil also slows down weed growth during establishment phases.

SNOW FENCES

Driving in snow is bad enough, but having to contend with deep, hard-packed drifts can make it truly nerve-wracking and potentially fatal. For rural dwellers, warming up the tractor and bundling up to plow snow or extract stuck motorists can be an unpleasant and bone-chilling task.

Snow movement can be controlled with fences designed to reduce wind speed and allow the heavier particles to come to rest in a drift away from the road, corral, or other area you need kept clear. Natural examples of this process are seen where a row of trees are close to the edge of the road or where a high bank of earth is beside a road cut into a hill. In either case, the wind slows down and the snow drifts in.

Using fences to manage snow movement is approximately 100 times less expensive than plowing snow off winter roads, according to research by the United States Strategic Highway Research Program (SHRP). Since this study was done a number of years ago, when fuel prices

A shelter belt adds beauty to a farmstead or acreage and enhances a property's value. Shelter belts help people live and work more comfortably by modifying the environment. They act as natural air conditioners in the summer and buffer bitter winds in the winter. They can save property owners money by reducing both winter heating and summer cooling requirements. A shelter belt can reduce your home energy consumption by up to 36 percent. Feed conversion in livestock protected by a shelter belt is greatly improved because of the reduction of wind chill. A well-designed shelter belt can prevent snow from drifting onto driveways and work areas and can act as an effective visual and noise barrier. In addition, shelter belts provide countless forms of wildlife with critical winter shelter and breeding habitat.

Iowa Department of Natural Resources, July 2001
www.iowadnr.com/wildlife/files/FMAjuly01.html

A snow fence can be attached to existing fences if they are far enough back from a road. The effectiveness of this fence would be improved by raising it to leave a gap underneath.

Strategically placed support for a snow fence can be quickly installed after crops have been harvested.

were much lower, the cost disadvantage of plowing snow can be expected to be even worse today. In terms of the inconvenience of getting out to plow snow, the fence is always on duty, so you don't face the unpleasant task of plowing when the school bus must get through or the livestock need feeding.

Several types of snow fences are available: the commonly used vertical-slat snow fence, the plastic-mesh snow fence, and the Wyoming-style snow fence. The latter is an 8-foot-tall slatted panel with a frame to hold it 10 inches off the ground and lean the panel into the wind about 15 degrees from vertical. Studies indicate that the Wyoming-style snow fence is the most efficient, followed closely by the plastic snow fence, and then the somewhat less efficient vertical-slat snow fence. A snow fence constructed of horizontal boards having 50 percent air porosity is 25 percent more efficient at managing snow than one constructed of vertical boards.

However, the practicalities of cost, convenience of installation, and durability also come into play. The Wyoming-style snow fence can be costly and labor intensive

Key Tips For Snow Fences

- The best fence porosity (ratio of solid surface to total surface) is 40 to 50 percent.
- The distance between fence and road should be at least 35 times the height of the fence. Snow fences too close to the road end up forming the drift right on the road, which can be handy if you want an access road closed for the winter.
- Although fences should be perpendicular to the prevailing wind direction, the angle can vary by as much as 25 degrees without affecting performance. It is useful to know where winds are not reliable.
- To maximize the effectiveness of any snow fence, high or low, leave a gap under the fence equal to 10 percent of the total fence height.
- For effectiveness and economy, a single row of tall fences is always preferable to multiple rows of shorter fences.
- One 6-foot fence is as effective as two rows of 4-foot fence.
- One 8-foot fence is as effective as five rows of 4-foot fence.

to build, but it may be your only choice for permanent installations where snow is a persistent problem. For durability's sake, the face of the Wyoming-fence is also slightly tilted so that the cattle cannot rub up against it and gradually knock it down.

Plastic-mesh snow fence has 50 percent porosity, and as a result is efficient, reasonably priced, and relatively easy to install. While a vertical wooden-slat fence is the least efficient at managing snow, it is also inexpensive and easy to install.

If your decision is to go with the lower-cost and easier-to-install fence, remember that you can optimize efficiency and longevity by raising the bottom of the fence slightly above the ground. The recommended lower gap is 10 to 15 percent of the total snow-fence height.

A strong, high hedgerow upwind from the road at far left helps keep it clear of snowdrifts in winter. When drifts melt in spring, the moisture benefits the crop between the road and hedgerow.

Stages of Snowdrift Formation

Some snow does pile up in front of the fence, but much more is trapped downwind. The fence's wind energy reduction zone, which extends about 15 times the height of the fence, causes a lens-shaped ridge of snow particles to collect downwind of the fence. Due to air eddying at the downwind end of the lens, a slip-face begins to form, adding significant resistance to the approaching wind. The eddy zone and slip face help trap particles blowing off the top of the drift. The drift becomes deeper, but not much longer.

The eddy zone fills in as the drift lengthens downwind. Snow-trapping efficiency declines as the eddy zone diminishes. The drift profiles then smooth out as the eddy zone disappears. The drift is now about twenty times the height of the fence, and slowly grows to its final length of thirty to thirty-five times the height of the fence. This limit is why the fence should be placed that far upwind of the roadway to be protected.

For more detailed information, adaptable to most climates, you can also refer to the University of Minnesota interactive snow fence design tool to help you plan where to locate your fence: www.climate.umn.edu/snow_fence/Components/Design/introduction.htm.

MOISTURE MANAGEMENT

While a snowdrift presents a lot of bother when it winds up in an inconvenient place, such as on a road, snowdrifts can also provide a useful moisture resource when they melt. Snow is nowhere near as good as rain in terms of a moisture resource, as it takes about 4 to 7 inches of melted snow to provide as much moisture as 1 inch of rain. But in drier regions where farmers welcome every drop of growing season moisture they can get, trapping snow provides a useful gain.

Water levels can be raised in ponds or dugouts by placing snow fences in areas to cause snowdrifts to form on the water/ice surface itself. If the area around the pond has well-established grass, snow fences can also be placed to create snowdrifts on any slopes that will drain into the pond in the spring. Placing snow fencing around a drainage basin may not be a good idea where the

DOT Encourages Farmers to Participate In Snow Fence Program

Ames, Iowa—While many Iowa farmers are thinking about the fall harvest, the Iowa Department of Transportation maintenance officials would like to encourage landowners to include participation in the DOT's snow fence program in their harvest plans.

Through Iowa's Cooperative Snow Fence Program, the Iowa DOT works with private landowners to create snow fences along roadsides. As part of the ongoing effort to meet the challenges of winter maintenance on Iowa roadways, the DOT would like to expand this program and encourage landowners to increase their participation. The fences can be permanent or temporary, use wood and plastic, or they can be living structures like trees, shrubs, corn stalks, or tall grasses.

According to DOT officials, one of the most cost-effective types of snow barriers is rows of corn planted parallel to the roadway in selected locations. Some sections of the corn are left standing after harvest and serve to trap snow before it reaches the roadway. The snow fence program does provide for payment to farmers for the unpicked corn if it is in a location where a barrier is needed. Farmers will be paid for the corn, but are also able to either hand harvest in the fall or use a corn picker in the spring to remove the crop. Leaving corn stubble in the fields through the winter and disking in the spring instead of the fall is also helpful.

Landowners and the public benefit by having snow fences on private property. Landowners can improve winter access to their farmsteads and other rural areas, help reduce soil erosion, provide habitat for wildlife, and increase crop yields by retaining moisture and reducing the drying effects of the wind.

The benefits to the public from having less blowing and drifting snow on the roadways include lower accident rates, increased visibility, decreased freezing and thawing effects on the roadways, and reduced need for constant snow removal. The national Strategic Highway Research Program has shown that it costs significantly more to plow snow than to trap it with a snow fence.

ground is prone to erosion from running water. You may wind up with gullies cut into the land and a pond full of muddy water.

Hedgerows aligned more or less perpendicular to prevailing winter winds are a permanent fence for trapping snow in fields and pastures. Since winter winds tend to come from a northerly direction, the snow accumulates on the southerly (downwind) side of the hedgerow. As the heat of the early spring sun intensifies from the south, the drifted snow melts rapidly to provide soil moisture for the growth of plants. The same effect of trapping snow can be achieved by erecting snow fences. The effect may not be as large because the fence is not as tall, but it will still be useful.

In the northern Great Plains and Canadian prairies, many farmers leave the stalks of cut crops standing all winter. In the reduced-tillage or zero-till farming systems used by such growers, the stalks act as a sort of mini–snow fence. While the height is not large, the extent is huge, so useful amounts of moisture can be accumulated for use by spring-planted crops. Leaving taller stubble also helps maintain adequate snow cover on fall-planted crops, such as winter wheat. The increased cover reduces winter kill and improves soil moisture levels for spring growth. There may also be cash benefits in using crops for snow fencing, as outlined in the Iowa Department of Transportation (DOT) press release in the sidebar. Check to see if your state has a similar program.

ANTI-EROSION FENCES

Anti-erosion fences are most commonly associated with gullies (see photos on page 156). Rushing water can result in rapid and severe soil degradation and associated ill effects downstream. Placement of a low and strongly supported barrier slows down the water and reduces its ability to carve into and carry off soil.

If you are a small farmer relatively new to the practice of growing field crops, consider the value of standing stubble to reduce soil erosion and degradation. Whether or not you have snow, the standing stalks provide a ground-level microclimate that helps keep the soil from drying out over winter and therefore less prone to erosion from wind. A thin but undisturbed layer of chaff and chopped stalks left on the soil surface also helps in this regard.

An even greater benefit comes from helping prevent erosion by moving water, which causes much more rapid and deep soil erosion than wind. During fall or spring rains, the water must either infiltrate the soil or flow somewhere. In general, the less land is disturbed by tillage, the easier it is for water to infiltrate, and the less damage water causes if it does flow across. For example, a healthy, well-established pasture suffers much less damage from water erosion compared to a bare field.

When crop stubble is left standing instead of being plowed down after harvest, it helps anchor the soil, similar to (but not as strongly as) the way grass anchors the soil in a pasture. By not plowing the soil after harvest, soil structure also remains so that water can more easily infiltrate. For fields in long-term reduced-tillage farming, a big part of this improved structure comes from the greatly increased area of earthworm burrows.

Plowing may improve water infiltration for a short period of time after an initial tillage operation. But a large body of scientific research and farmer experience shows that the long-term result of plowing is generally a steady decline in soil porosity and permeability, along with crucial losses

(Above) Water can also be managed with fences, such as this barrier to prevent erosion and help get grass established in a ditch bottom. *(Right)* On the other side of the road, where there is no erosion barrier, the consequences are devastating.

in soil organic matter and the associated increases in soil carbon emissions.

Soil permeability (the natural capacity of soil to conduct water) depends on the amount and stability of soil pores. Both are enhanced by practicing a system of farming that leaves crop stubble standing over the winter. Both are degraded by plowing. Your crop stubble is therefore a plain-looking but very important anti-erosion fence.

For more information on the part played by crop stubble in erosion and soil improvement, please review as much information as you can on reduced-tillage farming. Depending on where you live, such practices go by other names, including min-till, no-till, or zero-till. Although it is often used in conjunction with nonselective herbicides to control weeds, organic reduced tillage is also possible and is being actively researched on methods of improvement. A few sources for getting started on learning about reduced tillage are listed in Resources, under Environmental Management on page 169.

A small creek runs through this pasture, and rather that tearing up the creek banks with ATV and foot traffic, this owner has built a sturdy bridge entirely out of fencing materials such as posts and planks. To keep vehicles from rattling across the deck of the bridge, a coating of road gravel has been laid on the bridge deck.

CHAPTER

10

REMOVING FENCES

Through the passage of time or changes in land use, it may become necessary to remove fences that are no longer functional or needed. Before you start pulling posts and coiling wire, review the issues involved so that you don't create legal or safety issues for yourself.

Fences wholly within your property are your own business, but boundary fences involve the interests of adjacent property owners, so neighbors should be consulted before you remove or modify any boundary fence. In fact, the law in most jurisdictions requires something along the lines of "No boundary fence is to be removed without the consent of all interested parties." Even if the fence had mistakenly been constructed inside the boundaries of your land, in which case you are usually within your rights to

Fence Removal and Adverse Possession

If you're acquiring or operating a small farm in an area without recent land surveys, you may encounter a situation where a long-established custom conflicts with your right to remove (or build) a fence. The summary of this Connecticut case illustrates the typical principle.

In 1970, 13-year-old Kimberly Johnson and her 11-year-old brother erected a wire fence to enclose Kimberly's new horse. The children strung three parallel wires on existing trees, believing they were stringing the wire fence entirely on their mother's property.

Horses and/or burros were confined in this corral nearly continuously from 1970 to 1997. Beginning in 1985, the Johnsons upgraded sections of the wire fence with a wooden fence.

The wire fence, the wooden fence, the paddock area, and the animals were conspicuous and easily seen from most perspectives. Their neighboring landowners were well aware of its existence, location, and use.

In 1994, the neighboring property was sold to the Kelloggs (plaintiff). Before the purchase, the Kelloggs walked the property and were aware of the existence of the fenced enclosure and barn. They did not have a legal survey at the time of purchase and assumed at that time that the enclosure was on the Johnson property.

In 1999, the Kelloggs hired a licensed land surveyor to stake out the property boundary so that they could build a boundary fence. The surveyor found that the Johnsons' old fence encroached on the Kelloggs' property.

On learning of the encroachment, the plaintiffs requested that the defendants remove the old fence. The defendants declined and a lawsuit ensued.

The trial court ruled that the defendants acquired legal ownership of the disputed parcel by adverse possession. They occupied it openly and without seeking permission from the recorded owner for the required time (15 years in this jurisdiction). Accordingly, they, not the defendants, own the disputed parcel.

Had the Kelloggs simply gone ahead and removed the offending fence, they would have been in far worse shape. They would have still lost the land, but also then probably would have been responsible for all the costs of replacing the fence.

Points from *Kellogg v. Johnson*, Connecticut Superior Court 2001 WL 1420608, October 26, 2001.

remove it, it may also be a requirement that you give some period of notice (e.g., one month) to neighboring landowners that you are going to remove the fence.

These are just a few simple legal issues that affect boundary-fence removal. Things become rapidly more complex if one owner wants the fence for livestock containment, while the other wants the fence gone for aesthetic reasons, or if you encounter issues of long-held customary use (see sidebar on page 159). There may be many more issues, and the only way to be sure is to review the actual laws and regulations in your jurisdiction. This doesn't necessarily need to involve complex legal consultation. Checking with a county clerk or the equivalent may be all you need to avoid stirring up legal hassles by removing a fence.

SAFETY ISSUES

Along with the cuts and scrapes that normally go along with handling fence wire, there is the much larger danger associated with pulling out posts. Use a proper lifting device such as a front-end loader post or post jack. Do not use improvised methods (see sidebar on next page).

When you're using a tractor and front-end loader to pull posts, do not pull posts with the tractor tilted a lot to one side. If resistance from the post causes the rear wheels to lift off the ground, the tractor could roll sideways. Where the ground is noticeably sloped, position the tractor so that it is downhill or uphill from the post.

Fingers can be caught and crushed in chains used to pull posts. When attaching chains, wear heavy work gloves and do not give a signal to lift the post until your hands are well clear of the chain.

When the bucket of a tractor's front-end loader is lowered to hook up to the post for pulling, persons standing underneath can be struck or crushed. When you give a signal to lower the bucket, watch the bucket and not the post. Make sure the tractor operator is trained in understanding your signals and in smooth movement of the front-end-loader hydraulic controls.

If a post strongly resists extraction and/or suddenly releases, there can be unpredictable effects on the stability of the tractor being used to pull the post. Instead of trying to jerk it upward, try to wiggle the post from side to side to loosen it from the ground. If it's too tight to wiggle by hand, push on the base with the front-end loader.

Along with the safety of people removing the fence, consider the safety impacts on animals. For example, cattle are noted for eating metal objects that then puncture their stomachs (hardware disease). For this reason, do not fling removed staples on the ground. Gather them up and take them away, along with any small bits of wire. Other animals, especially sheep, can get entangled in coils of wire left lying around,which will cause cuts and stress.

When pulling out posts, pay close attention to safety issues. Along with tractor turnover risks, there are a lot of hazards for anyone working close to and underneath the front-end loader.

Teenager Dies in Tractor Overturn While Removing Fence Posts

A 14-year-old boy died while helping his grandfather clear farmland of posts from an old electric fence line. The boy was operating a utility tractor equipped with a homemade roll bar and seat belt. They had successfully removed several posts using the tractor's three-point hitch lift arms, but encountered a post that wouldn't budge. The grandfather switched to another method to pull posts using the rear wheel, something he had learned from his father many years ago. They backed up the tractor into position, adjacent to the 6-inch-diameter wooden fence post. The tractor was facing up a slight slope. A chain was secured to the bottom of the post and the other end was looped around the rear tire. Normally, as the wheel would turn, it would pull the post upward and forward. As the boy released the clutch the post did not move but the tractor's front end jumped upward. The grandfather yelled a warning, but the boy had no time to react, and the tractor rolled completely over backwards. The self-made roll bar was sturdy and did not itself break or bend. However, the home-modified brackets attaching the roll bar to the tractor did fail, snapping off large pieces of the axle housing as the roll bar collapsed against the tractor. The boy was instantly crushed underneath.

from Iowa FACE Report 03IA020

PRACTICAL ISSUES

If you're using a front-end-loader-equipped tractor to pull posts, add some weight at the rear of the tractor to resist tipping. A ballast box or heavy implement on the three-point hitch is often enough. If you find the chain is sliding off when you try to pull the post up, try loosely knotting the chain around the post. This usually makes the chain dig into the wood and improves grip.

Remove the wire before pulling posts. Trying to roll up a fence with wires still attached to the posts quickly becomes an overwhelming mess. A tractor-mounted hydraulic wire winder is very handy for simplifying and speeding up the messy job of winding up removed wire.

Old fence wire is difficult to dispose of, because metal recyclers often don't want to accept it. Plan on a long-term storage space where animals won't get themselves caught in it. Also plan out a space to store long-term and treated fence posts you pull out. Do not burn any kind of treated wood, including fence posts, because it can produce toxic smoke and ash. The ash tastes salty to animals, and that could lead to poisoning of your own livestock if they eat the ash from the post you burn.

A sleeve is crimped on to secure the loop of wire that goes around the post.

CHAPTER

11

PAST AND FUTURE FENCES

Stone fences have an ancient history in the Old World, and early colonists in North America used stones from their fields to build livestock fences.

In the previous chapters, we've seen that fence designs are always changing as new technology becomes available to fill the key need of low-cost, low-maintenance movement control of livestock, predators, and pests. One of the most dramatic changes was the adoption of barbed-wire fencing. But even well before barbed wire, farmers were testing innovations with rail fencing to make it less costly in terms of labor and land use. In more recent times, there has been a steady move toward high-tensile smooth wire fencing and various kinds of electric fencing.

There are already signs that the next big innovation in fencing may involve guidance with radio signals and GPS location. Instead of fixed fence lines on the ground that provide physical and/or pain barriers, the idea is to provide electronically generated cues that prompt animals to move or stop. The technology already exists for pets in the form of "invisible fences" for yards. Tracking collars for wildlife (and some low-security human prisoners) are another part of the puzzle that's already in wide use.

The biggest current barrier to the "virtual fence" is cost. Use of invisible fences is practical for pets because the owner is only dealing with a few animals at most, aesthetics are highly important, and pets have a high perceived value. Tracking collars for wildlife solve the need for a noninvasive method and only track the animal without influencing its movement. For a livestock manager dealing with hundreds of animals and faced with profits of only a few cents a pound on the animal products, the technology of electronic fences can't help but appear to be too expensive and not sufficiently refined enough to be practical.

THE EVOLUTION OF FENCING

It's funny how things change as seemingly unrelated external trends grind along. Shortage and abundance are key drivers in fence design. Farmers managing livestock have always faced the development of shortages that hampered existing systems. For example, the farmers in colonial America were used to enclosing their fields

As settlers moved west, they found that stones weren't as abundant. To build fences, they turned to the most readily available material—wood.

with walls made from stones that were turned up from the soil by cultivation. But for farmers in the larger, less stony fields out west in Ohio and Illinois, the shortage of stone led to innovation in the use of what was abundant: wood. Split-rail "snake" fences were one of the cheap, effective solutions developed. As farmers ventured farther west onto the treeless Great Plains, and as settled areas became cleared of trees, fence builders were once again faced with a shortage of the favored material–wood–and had to again innovate in response.

Farmers were also facing another growing shortage by the late 1800s: labor. Industrialization was drawing off surplus and not-so-surplus labor from the countryside. Why work at splitting rails or work the hard, lonely life of a cowboy when you could make good, steady money at the new factory jobs? Fortunately for farmers, one of the products of those new factories was enough cheap, abundant steel wire to make barbed-wire fencing a practical proposition. Steel was still a relatively rare and expensive material in 1867, but with advances in steel-making technology, global production increased over twenty-fold by 1870, doubled from that amount by 1880, and by the end of the century had increased over twenty-eight times. Note that these dates correspond pretty closely with the rapid spread of barbed-steel wire as a practical fencing material.

Barbed wire didn't have instant universal acceptance, however. Read anything related to the Old West cattle culture, and you'll quickly encounter tales of the Range Wars, where old-time ranchers

Farther west, on the Great Plains, wood in turn became scare. Many farmers and ranchers adopted barbed wire, a cheap, effective fencing material.

bitterly opposed the closing of the range with this newfangled "devil's rope" on moral, as well as practical, grounds. But some big ranchers, who were the largest opponents of barbed wire, quickly adopted it once they started improving their herds to meet the demand for better-quality beef. Fences helped keep lower-quality stock out and reduced the costs of hiring increasingly harder-to-find line riders.

At the other end of the economic scale, small farmers embraced barbed wire as a cheap, effective way to protect their crops and herds. In the Spanish-American areas of the West, reliable fences reduced the time and labor costs of the Mesta councils needed to identify and round up stock grazing on the old commons.

There were failures in this new fencing system. The "big die-up" in the winters of 1885 through 1887 resulted from cattle piling up against the unmonitored drift fences erected across the Texas panhandle. Cattle moving south to flee winter storms couldn't cross the fence, and wouldn't by nature turn back into the storm. Untold thousands of animals died, and many ranches failed.

But the failure led to innovation and improvement in barbed-wire fencing, not its abandonment. Similarly, many early attempts at electric fences failed because of inadequate energizers and fencing techniques. However, the failures drove innovations and improvement to the point where the electric fence, like barbed wire and high-tensile smooth wire, has become an accepted and familiar form of fencing today. At the individual farm level, ongoing failures from livestock escape and predator intrusion also drive innovation in fence upgrading and improvement.

As for current steel supply, global production is still high, but recent demand in East Asia has emerged as a huge market factor. Steel prices are rising so frequently that some farm machinery manufacturers are adding a steel surcharge to the cost of their equipment. Projects like petroleum pipelines must book steel-pipe production earlier and earlier in order to get enough pipe. If there's a worldwide shortage of steel, you can be pretty sure the effect is going to eventually be reflected in the supply of steel wire for agricultural fencing.

Electricity is another commodity whose price has become much more volatile lately. Electric fencing really took off in the low-cost-electricity environment of the late twentieth century, when some rural electricity rates were regulated to give farmers a price break. But as anyone can attest from personal experience, today's energy bills are substantially higher. Although today's electric-fence energizers are vastly more effective and energy efficient than those of even a few years ago, they still use a commodity that's getting more and more expensive.

VIRTUAL FENCING

Fence designs are refined as materials, money, time, and/or labor become scarce. For completely new designs to appear, it requires some new alternative to become cheap and abundant. In the 1880s, it was steel for barbed wire. In the 1970s, it was high-tensile smooth wire. Today, what are clearly becoming cheaper and abundant are communications systems and computing power, which are being brought to bear on fence technology.

For example, at the Jornada Experimental Range near Las Cruces, New Mexico, animal research scientist Dean M. Anderson rounds up cattle with the help of global positioning system (GPS)

(Right) Researchers are pictured with a prototype virtual fence device that provides audio cues telling cattle which way to move. Virtual fencing offers a way to manipulate animal distribution without the need for wire and posts. *Scott Bauer, USDA-ARS*

(Below) The prototype virtual fence receiver is being "installed" on a cow. It's a bit of a struggle, but less trouble than building miles of fence. *Scott Bauer, USDA-ARS*

signals coming from satellites. In this United States Department of Agriculture–Agricultural Research Service project, cows are equipped with locator/controller collars that send electronic versions of the cowboy's movement commands into the cow's ears. The whispered commands act as a virtual fence.

Anderson is a longtime student of using cattle's innate behaviors to manage their movements in low-stress ways. His virtual fence uses electronic cues instead of a person to move cattle according to ecologically and economically sound grazing principles. Virtual fencing could soon offer a tool to improve grazing without the costs of close herding or conventional wire and posts. While he doesn't see an end to conventional boundary fences that protect safety and property rights, he believes virtual fences may be economically and environmentally judicious for management within property boundaries.

A CHANGING WORLD

In many areas of the world, conventional fencing is impractical and/or not economical. In the vast and ecologically vulnerable ranges of the West, fencing costs can be huge and even worse when fencing more rugged terrain. At another end of the same scale are the growing number of small farms in North America where the time needed for earning off-farm income can sometimes lead to a shortage of time for building and maintaining effective fences. Yet in these and many other cases, effective control of animal movement is desperately needed to earn a profit and prevent improper resource use.

A bold step, such as the adoption of virtual fences, may be just what's needed to resolve these intractable demands. It's similar to the kind of innovative steps taken by New Zealand farmers in order to make smooth, high-tensile wire fencing a practical reality. This once troublesome and scoffed-at fencing system is now a technology that has caught on worldwide and is being further perfected everywhere it has spread.

Can virtual fencing do the same? Scientists like Dean Anderson and others are doing their part in development. It will soon, as always, be up to individual farmers to decide whether the costs, labor savings, animal husbandry improvements, and ecological benefits will make virtual fencing another tool for practical farming.

For more information on virtual fencing at the Jornada project, see "The Cyber Cow Whisperer and his Virtual Fence," as published in *Agricultural Research* magazine (November 2000), available at www.ars.usda.gov. You can also obtain the article by calling (202) 512-1800 between 7:30 a.m. and 4:30 p.m., EST, or by writing to *Agricultural Research*, P.O. Box 371954, Pittsburgh, PA 15250-7954.

The prototype virtual fence device in its neck saddle is ready to give signals to the cow. Future versions are expected to be the size of an ear tag or even smaller. *Scott Bauer, USDA-ARS*

RESOURCES

COSTS OF FENCING

Costs of Cattle Fencing for Grazing Areas
University of Nebraska–Lincoln
www.ianrpubs.unl.edu/sendIt/ec830.pdf

Estimating Beef Cattle Fencing Costs
British Columbia Ministry of Agriculture, Food, and Fisheries
www.agf.gov.bc.ca/resmgmt/publist/300series/307050-4.pdf

Estimating Costs for Livestock Fencing
Iowa State University
www.extension.iastate.edu/Publications/FM1855.pdf

Fence Budgets
Virginia Cooperative Extension
www.ext.vt.edu/news/periodicals/fmu/1997-10/fence$$.html

Fencing Cost Calculator (interactive)
University of Kentucky
www.bae.uky.edu/ext/livestock/calculators.htm

Fencing Costs
Purdue University
www.agry.purdue.edu/ext/forages/rotational/fencing/fencing_costs.html

Fencing
Colorado State University
www.coopext.colostate.edu/chaffee/fences.html

Fence Planning and Estimating Worksheet
British Columbia Ministry of Agriculture, Food, and Fisheries
www.agf.gov.bc.ca/resmgmt/publist/300series/307050-2.pdf

Material Costs for One Mile of Various Types of Fencing
University of Florida
alachua.ifas.ufl.edu/livestock/fence.htm#materials

Permanent Fencing Costs for Cattle and Sheep
University of Minnesota
www.extension.umn.edu/Beef/components/homestudy/plesson3a.pdf

ELECTRIC FENCING: ENERGIZERS

Choosing an Energizer (Fence Charger)
University of Kentucky
www.bae.uky.edu/%7Elturner/charger.htm

Electric Fence: Volts, Joules, and Deep-Cycle Batteries
West Virginia University
www.wvu.edu/%7Eagexten/farmman2/shepwool/elefence.htm

Electric Fence Controllers
British Columbia Ministry of Agriculture, Food, and Fisheries
www.agf.gov.bc.ca/resmgmt/publist/300series/307310-1.pdf
Electric Fence Energizers
Iowa Beef Center
www.iowabeefcenter.org/pdfs/bch/06201.pdf

Electric Fence Energizers
University of California–Davis
www.foothill.net/%7Eringram/energzer.htm

ELECTRIC FENCING: GENERAL TOPICS

Seventeen Mistakes to Avoid with Electric Fencing
Sustainable Farming Connection
www.ibiblio.org/farming-connection/grazing/features/fencemis.htm

Electric Fence Design
University of California–Davis
www.foothill.net/%7Eringram/design.htm

Fencing Management: Electric Fencing
Ontario Ministry of Agriculture, Food, and Rural Affairs
www.omafra.gov.on.ca/english/crops/pub19/5elecfen.htm

Electric Fencing for Sheep
Oklahoma State University
www.osuextra.okstate.edu/pdfs/F-3855web.pdf

Everything You Need to Know About Electric Fences
Manitoba Agriculture, Food, and Rural Initiatives
www.gov.mb.ca/agriculture/livestock/beef/baa10s01.html

Introduction to Electric Fencing
British Columbia Ministry of Agriculture, Food, and Fisheries
www.agf.gov.bc.ca/resmgmt/publist/300series/307300-1.pdf

Permanent Electric Fence Materials
Iowa Beef Center
www.iowabeefcenter.org/pdfs/bch/06202.pdf

Portable Electric Fence Materials
Iowa Beef Center
www.iowabeefcenter.org/pdfs/bch/06203.pdf

Temporary Electric Fence Materials Evaluation
University of California–Davis
www.foothill.net/~ringram/tmpfence.htm

Training Livestock to Electric Fences
University of California–Davis
www.foothill.net/~ringram/training.htm

Training, Testing, and Troubleshooting
British Columbia Ministry of Agriculture, Food, and Fisheries
www.agf.gov.bc.ca/resmgmt/publist/300series/307320-2.pdf

ELECTRIC FENCING: GROUNDING

Make a Well-Grounded (Earthed) Fence
Sustainable Farming Connection
www.ibiblio.org/farming-connection/grazing/features/ground.htm

Grounding Electric Fences
University of California–Davis
www.foothill.net/~ringram/groundng.htm

Grounding Systems for Electric Fences
British Columbia Ministry of Agriculture, Food, and Fisheries
www.agf.gov.bc.ca/resmgmt/publist/300series/307320-1.pdf

ENVIRONMENTAL MANAGEMENT

Field Windbreaks
University of Nebraska–Lincoln
ianrpubs.unl.edu/Forestry/ec1778.htm

How Windbreaks Work
University of Nebraska–Lincoln
ianrpubs.unl.edu/forestry/ec1763.pdf

Iowa's Cooperative Snow Fence Program
Iowa Department of Transportation
www.dot.state.ia.us/maintenance/pdf/snowfencebooklet.pdf

Living Snow Fences
University of Nebraska–Lincoln
www.unl.edu/nac/aug94/snowfences/snowfence.html

Snow Fence Guide
U.S. Strategic Highway Research Program
www.trb.org/publications/shrp/SHRP-H-320.pdf.

Wind and Snow Control Around the Farm
Purdue University
www.ces.purdue.edu/extmedia/NCR/NCR-191.html

Windbreak Design
University of Nebraska–Lincoln
www.ianrpubs.unl.edu/epublic/pages/publicationD.jsp?publicationId=467

Windbreak Establishment
University of Nebraska–Lincoln
ianrpubs.unl.edu/forestry/ec1764.htm

Windbreak Management
University of Nebraska–Lincoln
ianrpubs.unl.edu/forestry/ec1768.htm

Windbreak Renovation
University of Nebraska–Lincoln
www.unl.edu/nac/brochures/ec1777/index.html

Windbreaks for Livestock Operations
University of Nebraska–Lincoln
www.ianrpubs.unl.edu/sendIt/ec1767.pdf

Windbreaks for Snow Management
University of Nebraska–Lincoln
ianrpubs.unl.edu/forestry/ec1770.htm

Windbreaks and Wildlife
University of Nebraska–Lincoln
http://ianrpubs.unl.edu/forestry/ec1771.htm

HIGH-TENSILE WIRE FENCING

A Simple Wire Tension Meter
University of California–Davis
www.foothill.net/~ringram/tension.htm

Construction of High-Tensile-Wire Fences
University of Florida
edis.ifas.ufl.edu/AE017

Splices for High-Tensile, Smooth Fencing Wire
British Columbia Ministry of Agriculture, Food, and Fisheries
www.agf.gov.bc.ca/resmgmt/publist/300series/307131-1.pdf

Working with High-Tensile-Fence Wire
University of California–Davis
www.foothill.net/~ringram/hitensle.htm

GENERAL INFORMATION

Beef Home Study Course–Fencing System
University of Minnesota Extension
www.extension.umn.edu/beef/components/homestudy/plesson3a.pdf

Commonly Used Wire for Agricultural Fences
British Columbia Ministry of Agriculture, Food, and Fisheries
www.agf.gov.bc.ca/resmgmt/publist/300series/307100-1.pdf

Constructing Wire Fences
University of Missouri Extension Service
www.muextension.missouri.edu/xplor/agguides/agengin/g01192.htm

Fence Construction Safety
British Columbia Ministry of Agriculture, Food, and Fisheries
www.agf.gov.bc.ca/resmgmt/publist/300series/307050-3.pdf

Fence Planning for Horses
Penn State University
pubs.cas.psu.edu/freepubs/pdfs/ub037.pdf

Fence Posts: Materials, Installation, and Removal
British Columbia Ministry of Agriculture, Food, and Fisheries
www.agf.gov.bc.ca/resmgmt/publist/300series/307110-1.pdf

Fence Wire: Dispensing, Stapling, Joining, Tying
British Columbia Ministry of Agriculture, Food, and Fisheries
www.agf.gov.bc.ca/resmgmt/publist/300series/307100-2.pdf

Fences for the Farm
University of Georgia
www.pubs.caes.uga.edu/caespubs/pubcd/c774.htm

Fencing Factsheet
Ontario Ministry of Agriculture, Food, and Rural Affairs
www.omafra.gov.on.ca/english/engineer/facts/99-057.htm

Fencing Materials for Livestock Systems
Virginia Cooperative Extension
www.ext.vt.edu/pubs/bse/442-131/442-131.html

Fencing for Bison
Saskatchewan Agriculture and Food
www.agriculture.gov.sk.ca

Fencing for Goats
Langston University
www.luresext.edu/goats/library/field/hart01.html

Gates, Cattle Guards, and Passageways
British Columbia Ministry of Agriculture, Food, and Fisheries
www.agf.gov.bc.ca/resmgmt/publist/300series/307400-1.pdf

Hinged-Wire Gate
British Columbia Ministry of Agriculture, Food, and Fisheries
www.agf.gov.bc.ca/resmgmt/publist/300series/307410-1.pdf

Lightning Protection for Farms: Wire Fence Grounding
National Ag Safety Database
www.cdc.gov/NASD/docs/d001801-d001900/d001825/d001825.html

Livestock Control: Nonelectric Fence Designs
British Columbia Ministry of Agriculture, Food, and Fisheries
www.agf.gov.bc.ca/resmgmt/publist/300series/307260-1.pdf

Livestock Control: Electric Fence Designs
British Columbia Ministry of Agriculture, Food, and Fisheries
www.agf.gov.bc.ca/resmgmt/publist/300series/307260-2.pdf

New Fence Construction
British Columbia Ministry of Agriculture, Food, and Fisheries
www.agf.gov.bc.ca/resmgmt/publist/300series/307050-1.pdf

Pasture Fencing for Horses
British Columbia Ministry of Agriculture, Food, and Fisheries
www.agf.gov.bc.ca/resmgmt/publist/300series/307260-3.pdf

Pasture Management Home Study: An Online Course on Fencing Systems
University of Maine Cooperative Extension
www.umaine.edu/umext/pasture/Lessons/L3/intro3.htm

Planning and Building Fences on the Farm
University of Tennessee Agricultural Extension Service
www.utextension.utk.edu/publications/pbfiles/PB1541.pdf

Selecting Wire Fencing Materials
University of Missouri Extension Service
www.muextension.missouri.edu/xplor/agguides/agengin/g01191.htm

Types of Fences, Planning, and Legislation
British Columbia Ministry of Agriculture, Food, and Fisheries
www.agf.gov.bc.ca/resmgmt/publist/300series/307050-1.pdf

Wire Fence Brace Assemblies
www.agf.gov.bc.ca/resmgmt/publist/300series/307220-1.pdf
www.agf.gov.bc.ca/resmgmt/publist/300series/307220-2.pdf

Wire Fence Construction
British Columbia Ministry of Agriculture, Food, and Fisheries
www.agf.gov.bc.ca/resmgmt/publist/300series/307600-1.pdf

Wood Fence Construction
British Columbia Ministry of Agriculture, Food, and Fisheries
www.agf.gov.bc.ca/resmgmt/publist/300series/307600-1.pdf

Wood Preservatives and Treated Wood
Washington Toxics Coalition
www. watoxics.org/healthy-living/healthy-homes-gardens-1/resources-treated-wood

PREDATOR AND PEST EXCLUSION

Bear Wise: Technical Note
Ontario Ministry of Natural Resources
bears.mnr.gov.on.ca/technote_fence_permanent.html

Black Bears in Massachusetts
MassWildlife
www.mass.gov/dfwele/dfw/dfw_bears.htm

Building an Electric Antipredator Fence
Oregon State University
extension.oregonstate.edu/catalog/html/pnw/pnw225

Deer Exclusion Fencing for Orchards and Vineyards Using Woven-Wire Fencing
British Columbia Ministry of Agriculture, Food, and Fisheries
www.agf.gov.bc.ca/resmgmt/publist/300series/307251-1.pdf

Elk Exclusion Using Electric Fencing
British Columbia Ministry of Agriculture, Food, and Fisheries
www.agf.gov.bc.ca/resmgmt/publist/300series/307252-1.pdf

Elk Exclusion Using Woven-Wire Fencing
British Columbia Ministry of Agriculture, Food, and Fisheries
www.agf.gov.bc.ca/resmgmt/publist/300series/307252-1.pdf

Fence for Deer Exclusion
USDA—National Wildlife Research Center
www.electrobraid.com/wildlife/reports/USDAMAY02.html

Fencing Options in Predator Control
Ontario Ministry of Agriculture, Food, and Rural Affairs
www.omafra.gov.on.ca/english/livestock/sheep/facts/02-053.htm

Prevention and Control of Rabbit Damage, publication G1526
University of Nebraska-Lincoln
www.ianrpubs.unl.edu

When Coyotes Become a Nuisance
Nova Scotia Department of Natural Resources
www.gov.ns.ca/natr/wildlife/nuisance/coyotes.htm

When Raccoons Become a Nuisance
Nova Scotia Department of Natural Resources
www.gov.ns.ca/natr/wildlife/nuisance/raccoons.htm

PROPERTY LAW

Ohio Line Fence Law
Ohio State University
ohioline.osu.edu/als-fact/1001.html

Understand Liability Issues Before Raising Livestock
Iowa State University Extension Acreage Living Newsletter
www.extension.iastate.edu/acreage/AL2004/aloctnov04.html#liability

Missouri Fencing and Boundary Laws
University of Missouri Extension
muextension.missouri.edu/explore/agguides/agecon/g00810.htm

Fencing Law
Iowa State University Extension Acreage Living Newsletter
www.extension.iastate.edu/acreage/AL1999/almay99.html#fencing%20law

Virginia Law for Farmers and Landowners
Virginia Cooperative Fencing
www.ext.vt.edu/news/periodicals/fmu/2005-06/virginialaw.html

Questions Most Frequently Asked About Land Surveying
Louisiana Society of Professional Surveyors
www.lsps.net/promote/pls_faq.htm

Having Your Land Surveyed
Minnesota Society of Professional Surveyors
www.mnsurveyor.com/survey/survey_a1.asp

ROTATIONAL GRAZING

Introduction to Paddock Design, Fencing, and Water Systems for Controlled Grazing
National Sustainable Agriculture Information Service (ATTRA)
www.attra.org/attra-pub/paddock.html

Planning Fencing Systems for Intensive Grazing Management
University of Kentucky
www.ca.uky.edu/agc/pubs/id/id74/id74.htm

A Cattle Whisperer's Secrets
Watersheds.org
www.watersheds.org/farm/beckyday2.htm

Planning Fencing Systems for Controlled Grazing
Virginia Cooperative Extension
www.ext.vt.edu/pubs/ageng/442-130/442-130.html

Temporary Fences for Rotational Grazing
University of Tennessee
www.utextension.utk.edu/publications/spfiles/sp399G.pdf

SUPPLIES

Cameo Fencing
23 Triangle Road
Hammond, NY 13646
800-822-5426
www.cameofencing.com

Carolina Wire Products, LLC
www.carolinawire.com

Dare Products Inc.
860 Betterly Road
Battle Creek, MI 49015
800-922-3273
www.dareproducts.com

Electrobraid Fence
236 Water Street, Box 19
Yarmouth Nova Scotia, Canada B5A 4P8
888-430-3330
www.electrobraid.com

Fi-Shock Online
69 N. Locust Street
Lititz, PA 17543
800-800-1819
www.fishock.com

Gallagher Animal Management Systems
130 W. 23rd Avenue
North Kansas City, MO 64116
800-531-5908
www.gallagherusa.com

Grazier System–Canter, L.C.
7555 N. Greenwich Road
Wichita, KS 67226-8254
877-744-6150
www.graziersystem.com

Kencove Farm Fence
344 Kendall Road
Blairsville, PA 15717
800-536-2683
www.kencove.com

Kiwi Fence Systems and Supplies
344 Kendall Road
Blairsville, PA 15717
724-459-6952
www.kiwifence.com

Max-Flex Fence Systems
800-356-5458
www.maxflex.com

Parmak Electric Fencers
Parker-McCrory Mfg. Co.
2000 Forest Avenue
Kansas City, MO 64108
816.221.2000
www.parmakusa.com

Premier Fencing
2031 300th Street
Washington, IA 52353
800-282-6631
www.premier1supplies.com

Qual Line Fence
801 S. Division Street
Waunakee, WI 53597
800-533-3623 (toll free in WI, MN, MI, IA, IN, IL)
www.quallinefence.com

Ranch Fence
1080 Broadway
San Jose, CA 95125
800-213-2539
www.ranchfence.com

Red Brand Fence
Keystone Steel & Wire Co.
7000 S.W. Adams Street
Peoria, IL 61641
800-447-6444
www.redbrand.com

Stafix Electric Fencing, Ltd.
U.S. contact: South West Power Fence
26321 Highway 281 N.
San Antonio, TX 78260
800-221-0178
www.swpowerfence.com
Canada contact: Kane Veterinary Supplies Ltd.
11204 186 Street
Edmonton, Alberta T5S 2W2
800-252-7547
www.kanevet.com

Stay-Tite Fencing Manufacturing Inc.
1409 Freiheit Road
New Braunfels, TX 78130
888-223-8322
www.staytitefence.com

System Fencing
4919 7th Line Eramosa, RR #4
Rockwood, Ontario, Canada N0B 2K0
800-461-3362
www.systemfence.com

Tractor Supply Company
200 Powell Place
Brentwood, TN 37027
877-872-7721
www.mytscstore.com

Tractor Supply Company
440 Fence Company
4381 S. Highway 377
Aubrey, TX 76227
800-440-5889
www.440fence.com

Waterford Corporation Technologies
800-525-4952
www.waterfordcorp.com/WFhome.html

West Virginia Split Rail
P.O. Box 9
Buckhannon, WV 26201
800-624-3110
www.wvsr.com

Zareba Systems
13705 26th Avenue N, Suite 102
Minneapolis, MN 55441
Phone: 763-551-1125
Fax: 763-509-7450
www.redsnapr.com

INDEX

Also available from Voyageur Press

Audubon Birdhouse Book

ISBN 978-0-7603-4220-6

How To Raise Chickens

ISBN 978-0-7603-4377-7

The Whole Goat Handbook

ISBN 978-0-7603-4236-7

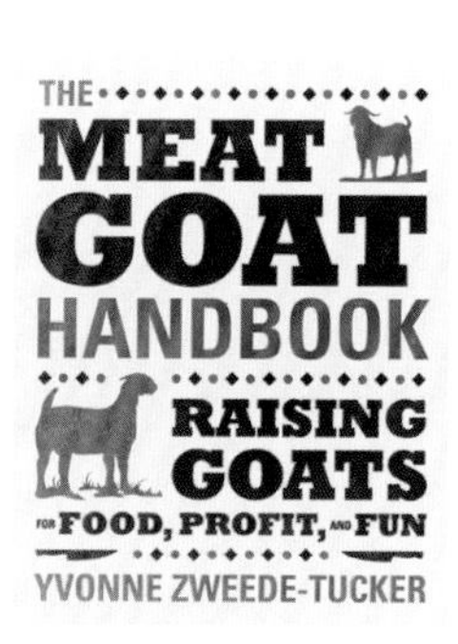

The Meat Goat Handbook

ISBN 978-0-7603-4042-4

The Beginner's Guide to Beekeeping

ISBN 978-0-7603-4447-7

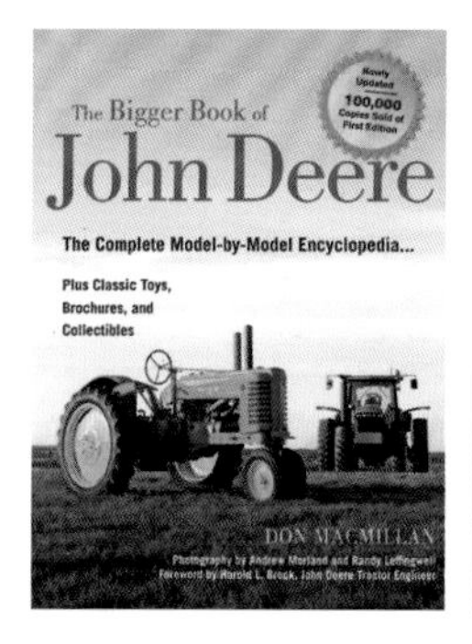

The Bigger Book of John Deere

ISBN 978-0-7603-4594-8

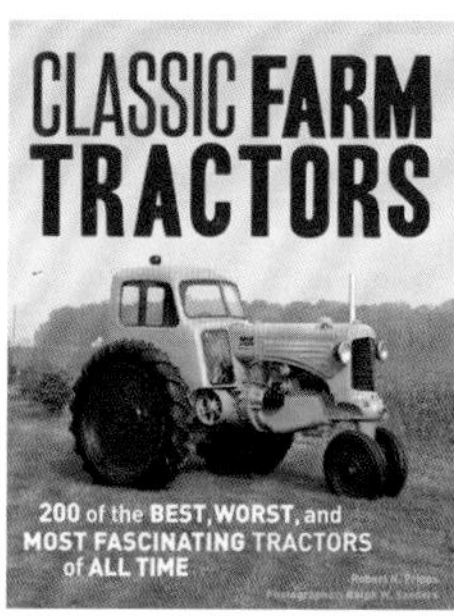

Classic Farm Tractors

ISBN 978-0-7603-4551-1

Driving Horses

ISBN 978-0-7603-4570-2

Homemade Cheese

ISBN 978-0-7603-3848-3

The Complete Book of Butchering, Smoking, Curing, and Sausage Making

ISBN 978-0-7603-3782-0